HOW TO
CREATE
YOUR
GARDEN

HOW TO
CREATE
YOUR
GARDEN

ADAM
FROST

CONTENTS

BUILD

ENJOY

GETTING STARTED

I love the way gardens help us connect to nature, the seasons, and the wider landscape. Whether you've got a tiny city oasis or an acre in the country, a well-thought-out garden is good for mind, body, and soul.

A lot of people are daunted by the idea of designing their own garden. With so much overwhelming information out there, it can seem a real challenge. However, in reality, garden design isn't that different from any other sort of design.

Think of it this way. Most of us can get our heads around turning our house into a home, choosing colors, buying furniture, and arranging it all so it meets our practical needs and feels comfortable, and garden design is not that different.

For me, a well-designed garden is essentially about four things: space, people, plants, and a sense of place.

The first step to great garden design involves understanding the space you are working with. By that, I mean the size, the soil, the climate, and the landscape.

Second, you need to understand that, above all else, gardens are about people, and the key to creating one that works well is to focus on what you want and need from it. Like your home, your garden should reflect your personality.

Third, think about plants. Many people have a big fear factor around plants, what with the Latin names and the idea that if something doesn't get pruned on the last Wednesday in March, the whole world will come to an end. In reality, learning about plants can be a journey that takes a lifetime, and you really don't need to know a lot to start.

Finally, if you can make some sort of connection to the area where you live, whether by growing particular plants or using local materials, then you'll end up with a stronger sense of place. Memories are a very important part of my design process—not just looking back on them but also looking at how we can create them for ourselves.

The garden should be a comfortable, usable space that reflects your needs and taste. Looking down from above, in its simplest form, a garden layout is just a series of shapes; how you balance those shapes impacts the way you use your garden and how it feels.

In this book, I'll show you how to develop your own garden style and choose the right plants for your garden area. I'll teach you the key principles of planting design, how to understand the way plants in nature work in layers, and how to mimic that in your own planting. I'll show you some simple building techniques and give you monthly tips and reminders to help you keep your garden looking beautiful from one year to the next. With a few insights and a step-by-step approach, I think that anyone can learn how to create a decent-looking, well-designed garden.

I want this book to get well worn, damp—even a little muddy! It will hopefully be a friend that will hold your hand while you are creating a space that is special to you. Remember, creating a garden is not about chasing perfection; it's about enjoying the journey and those little moments that surprise you along the way. I really believe that if you care about your garden, you will create a beautiful place that you want to be in.

Have fun!

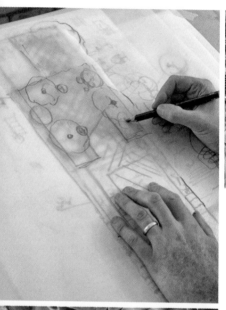

UNDERSTAND YOUR SPACE

DESIGN FOR PEOPLE

FIND YOUR STYLE

CHOOSE YOUR PLANTS

BRING IT ALL TOGETHER

DESIGN

DESIGN

UNDERSTAND
YOUR SPACE

> *Great gardens are all about
> emotions and how they make you feel.*

INTRODUCTION

The first stage of the design process is really all about information gathering, which will help you understand your space and what you want from it. Defining that at the outset will ensure that your design can realistically work within the space you have and stays true to your needs.

There are two ways to understand your space: the physical attributes—size, shape, average rainfall, soil, and so on—and the emotional—the way it makes you feel.

There's nothing quite like spending time in your garden area at different times of day and year to understand its emotional effect on you, and everyone's experience is personal to them, so a garden design needs to reflect that. Explore your space and find the places you instinctively want to linger with a cup of tea. Think about the views from the house and your surroundings.

I'll show you how to measure your garden area (which is not as scary as it might sound) and how to survey the soil, rainfall, and temperatures so you know what you have to work with. Then I'll ask you lots of questions to help you define what you want. This will all make sure the garden area you design is perfect for your requirements.

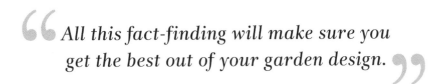

All this fact-finding will make sure you get the best out of your garden design.

GET TO KNOW YOUR SPACE

As you get ready to design your space, it's tempting to rip everything out and start with a blank canvas, but don't rush into anything like that just yet. First, you need to get a sense of your garden area's potential by observing what it currently looks like.

❶ WATCH AND LEARN

Spend some time in your garden area to really understand the space before you even start to consider embarking on any changes. Make notes on how the garden works at different times of the day. Repeat this regularly for at least a few months. Or, if you can, carry on for a whole year. This will give you a clear picture of how your garden changes across the seasons.

Every season can bring vivid color and interest to a garden.

Study your garden from inside as well as outside.

❷ TAKE IN THE VIEW

Look at your garden from every angle, including from the windows of your house, both upstairs and down. These views will form your strongest connection with the garden through the winter months and early spring when the weather keeps you indoors more. Start thinking about the views you enjoy and might want to emphasize, as well as those you'd prefer to hide (see pp32–33).

❸ LET THERE BE LIGHT

Watch how light and shadows fall and move around the garden area through the day (see p21). This will help you decide where various types of plants might be positioned as well as where you would ideally prefer to sit in the morning and evening.

Shady patches can be exploited to catch dappled sunlight.

Let your instincts lead you to places you feel naturally drawn to. Be aware of any emotional responses that may arise as you pause and contemplate.

6 BE MINDFUL

Note-taking shouldn't only be about hard facts and figures. As you spend time in the space, you may become aware that certain areas are simply nicer to hang out in than others. Some places will make you feel safe and relaxed, while in others you might feel overlooked and self-conscious. And a garden area—even one with nothing in it—always feels different when you move away from the house. Jot down any emotional responses like this, and build them into your designs later.

5 WHAT LIES BENEATH

It's worth waiting to see what's already growing in your garden area; each season some hidden gems may emerge. You might want to dig out established plants you're not wild about, but keep in mind that they may have taken years to grow and can form a strong backbone for your garden, giving it real structure. Ultimately, too, if you keep existing plants—even if you move, prune, or reshape them—you'll be saving money.

Pay attention to birdsong and other atmospheric sounds in your garden.

4 LISTEN

It's easy to get so used to noises around us that we no longer actually hear them. Yet sounds are really important when designing your garden, whether they enhance the atmosphere or need to be muffled. As you get to know your space, take the time to really listen to your surroundings.

Bulbs are like little jewels that add a whole new layer of interest to a garden when they emerge.

MAP OUT YOUR SPACE

The next step is to measure and map out your garden area's size, shape, and layout, including existing elements, any slopes, and where the sunlight falls. Gardens are rarely the shape we think they are, so drawing up a detailed plan is essential. It's not difficult and will help you all the way through the design process.

YOU WILL NEED

- sketch pad
- pencil
- 100ft (30m) tape measure
- ruler
- compass
- line and pins

Having an accurate scale plan of your garden is critical if you want to design it yourself. (You don't have to keep things where they currently sit, but it's worth getting to grips with what's already there and how it might affect your design and budget.) However, there's no point making a grand plan that is not to scale, as the design and proportions won't work when you come to build it. It will also be impossible to order the correct quantities of materials, so making an accurate drawing will save you money, time,

MEASURE A **BASIC SPACE**

These simple steps use methods called triangulation and taking offsets to map out your space accurately. Make a rough sketch first and use it to log all your measurements. Then transfer them onto paper to make a precise scale plan (see pp18–19). If you have an established garden area with lots of plants or structures that make access to the boundary tricky, you might find it helpful to look online at an overhead view. The farther you get from the house, the less critical it is to get accurate measurements; close to the house, it is key.

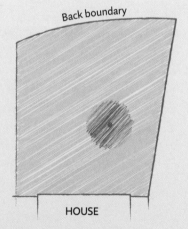

1 First, using your pencil and paper, roughly sketch the basic elements of your space: the house, the boundary line, and any fixed objects such as trees, buildings, structures, and manhole covers.

2 Measure from X across the plot to Y. Take measurements to mark the position of the house, windows, and doors. The two end points of the house are your fixed points A and B. Don't worry if the house isn't a straight line across the back; you need the corners only to provide the fixed points.

Measure across the back of the house and mark the positions of windows and doorways on your plan.

> *Gardens are rarely the shape we think they are, so drawing up a detailed plan is essential. It's not difficult and will help you all the way through the design process.*

and inconvenience. Bear in mind also that it's a lot easier (and cheaper) to change your plans on paper than it is to change a terrace or pathway once it's built. Most boundaries are not neat and square to the house, and they often have kinks, curves, or obstacles in the way, such as trees or structures, making measuring tricky. But don't worry. There are basic techniques, such as triangulation and taking offsets, that you can use to measure all manner of shapes and plot fixed features on your scale plan.

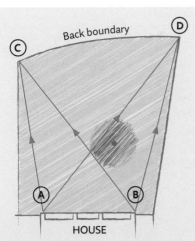

3 Now use triangulation so that you can accurately place the corners of your garden area when you create your scale plan. Measure from your fixed points A and B to the top left-hand boundary (C). Repeat for the top right-hand boundary (D). Note all measurements on your sketch.

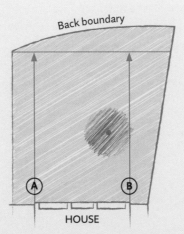

4 If parts of your garden are curved, you can take offsets to track the curve accurately. First, measure out two lines of exactly the same length, at 90° to the house, from A and B. Peg out a straight line at 90° to them so that it's parallel to the house.

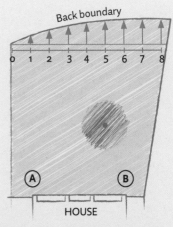

5 Along that pegged line, and at regular intervals, take measurements at 90° from the line to the perimeter to plot the curve. Make a note of those measurements, plus the distance of each interval, on your sketch.

MEASURE A **FIXED OBJECT**

To accurately mark a fixed object, such as a tree or shed, on your garden scale plan, you will need to take two measurements from your fixed points (A and B) to the point you wish to plot.

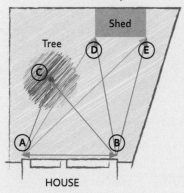

Note the distance from A to B on your plan. Now measure the distance from A to the fixed object, such as the tree (C), and note it down. Then measure from B to C and note it down. Repeat for other fixed points, such as the corners of your shed (D and E). Transfer these measurements to your scale plan to mark the location of fixed objects accurately.

MEASURE A **COMPLEX SPACE**

Triangulation is particularly useful if you have a garden with a more complex shape, such as a pronounced curve. The important thing is to have two fixed points from which to take your measurements. For greater accuracy, you may want to establish several fixed points on the house from which to work off. The more locations on the curve you measure, the more accurate it will be. If you have a big or complex garden, it may be worth investing in a professional survey, which can save time and money in the long run.

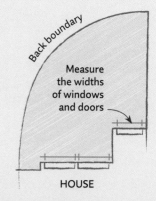

1 Make a rough sketch of the house walls that look on to the garden as well as the curved boundary. Measure and mark the size and position of windows and doors.

CREATE A **SCALE PLAN**

When you've taken all your measurements, transfer them onto paper to create a precise scale plan. First, figure out what scale will work best for your garden area. Most average-sized gardens are designed in either 1:50 (where 1in on your ruler = 3ft on your plan) or 1:10 (where ½in on your ruler = 3ft on your plan). If your garden is a bit larger, you may want to work at 1:100 (½in represents 39in). Use the longest measurement on your sketch to figure out the best scale to use. Using a scale ruler or a normal ruler, follow these steps to transfer the measurements onto paper. Lay tracing paper over your finished scale plan to sketch your design ideas for your plot. The scale plan is also great later when you need to figure out quantities for materials.

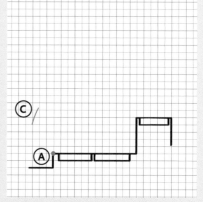

1 Transfer the measurements of the house, windows, and doors onto paper. Use a ruler to set your compass to the first scaled measurement you took from A to the boundary (C). For example, if the measurement is 16ft (5m) and you're using a 1:50 scale, you set your compass to 4in (10cm). Put the compass point on A and draw a small arc roughly where C is.

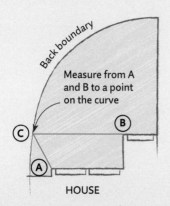

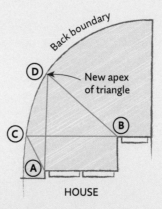

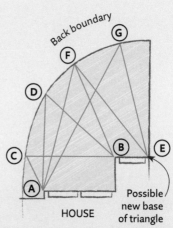

2 Choose two fixed points on the house for the base of your triangle, A and B. Measure between them, then measure from A to a point on the curve (C), then from B to C.

3 Now create another triangle by measuring from A and B to another point on the curve (D). Continue measuring in triangle shapes all round the curve.

4 Use new fixed points for some triangles if you need to (A and E). The more measurements you take, the more accurate your scale plan will be.

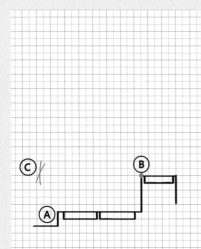

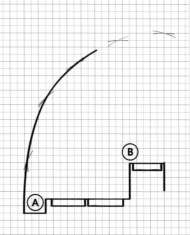

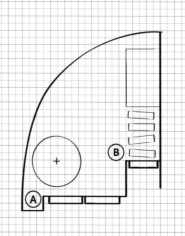

2 Reset the compass to the measurement you took from point B to C. Put the compass on point B and draw a small arc to cross the first arc. The point where the two arcs cross is the precise location of C.

3 Repeat this process for all the points you measured on the boundary. Then join the points up to create an accurate boundary line on your plan.

4 Use the same method to plot the position of other features in the garden, such as the corners or existing trees or a shed, to create your scale site plan.

SIMPLE **SLOPES**

Changes in levels can bring interest to a garden area, but they can also bring complications. They will have an impact on your design layout and the materials you use. If you've got a simple, small slope, it's fairly straightforward to figure out levels.

YOU WILL NEED

○ tape measure

○ hammer and wooden pegs

○ plank of wood slightly more than 3ft (1m) long

○ spirit level

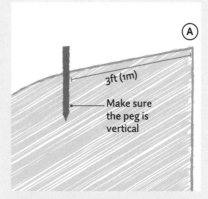

1 Measure 3ft (1m) down from the top of your slope (A). Hammer in a peg and use a spirit level to make sure the peg is vertical.

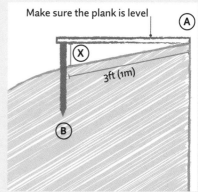

2 Lay the plank from A to the peg and use a spirit level to check that it's level. Hammer the peg further in as necessary until the plank sits horizontally. Measure the height of the peg (X).

COMPLICATED **SLOPES**

For more complicated level changes, I'd recommend renting some equipment, such as a laser-level device with a measuring staff, which allows you to get accurate readings around the garden. Don't panic—it's quite straightforward. How the device works can vary with the model, so make sure you read the instructions.

It's important to have a fixed point to put the laser level on, from which you take your readings, something that's not going to move from the point you start gathering information to the point at which the garden has been built (something such as a drain cover is ideal).

Next, take levels of the fixed elements that are going to stay in your design, such as a tree, a shed, or maybe the top of a wall. After that, if I feel that I still need more levels to understand the space, I knock in a series of pegs at different places around the garden.

Use the triangulation method (see pp18–19) to measure and note down the distance of those pegs from your two fixed points and add them to your plan. Then take height readings from the base of each peg, where it meets the ground, with the measuring staff and laser device. Transfer these measurements onto your scale plan.

Use a laser level with a laser detector attached to a measuring staff to take accurate measurements.

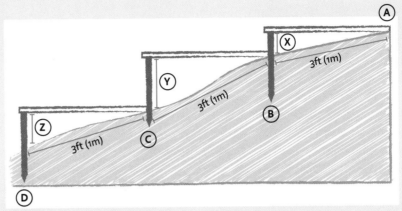

3 Hammer in another peg in 3ft (1m) further down the slope (C). Check that it's vertical. Lay the plank from the bottom of the first peg to the top of the second. Check that it's horizontal with the level. Hammer in the peg further as necessary.

4 Measure the height of the second peg (Y). Repeat these steps in 3ft (1m) intervals to the bottom of the slope.

5 To calculate the fall, or drop, of the slope, add the heights of the pegs together. The drop is the sum of the heights over the distance you have measured. Here, it's X + Y + Z over 9ft (3m).

FIND THE **SUNNY SPOTS**

Your garden's orientation (also called its aspect) will determine the amount of light and shade it gets at different times of the day and at various times of the year. This crucial piece of information will dictate the warmest places to sit, the best hard-landscaping materials to use, and the type of plants you can grow, as different plants prefer different levels of light and heat. To map out your garden's orientation, use a compass to locate north and mark this on your plan. Then mark out the sun's trajectory and annotate the plan with your notes on where the shade is and where the suntraps are.

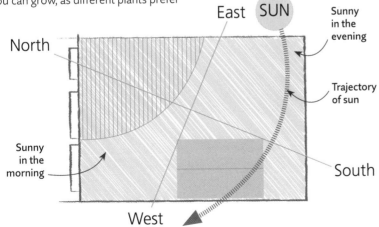

WORKING WITH
SPACE

This contemporary garden was designed as a space in which relatively large groups of people could socialize and garden together. The planting palette was chosen to sit comfortably with the materials, and the arching structure echoes the curve of the cedarwood path that pulls you slowly through the space.

Water and walls not only create great reflective surfaces but also cause you to pause as you move through the space. In the summer, the pool invites you to dangle your feet in or paddle.

Strong plant forms used through the space create rhythm and help lead the eye. I also used the same plant for the hedge to help bring the garden together.

I used matching materials on the surfaces and on the building. Using the wood and concrete in this way brought continuity to the garden.

GET TO KNOW YOUR SOIL

Soil is the single most important thing you've got in the garden, yet it's easy to take it for granted. You'll need to test it to figure out how it varies around the garden area, because the type of soil you have determines the range of plants you can grow. Luckily, this is quite easy to do.

When I'm teaching, it always amazes me how few people test their garden's soil. Yet we all agree that soil is fundamental to our gardens because it determines the range of plants you can grow as well as how successful they will be. You can figure out what sort of soil you've got largely by getting your hands into it, feeling the texture, and looking at the colour and composition. Knowing your soil's pH will allow you to grow a suitable range of plants, so I recommend buying a simple soil-testing kit (see p26). These are available from garden centers or nurseries, don't cost much, and are easy to use. Another thing to bear in mind is that the healthier the soil, the more resilient the plants will be when it comes to pests and disease. But you don't need to worry if your soil isn't in tip-top condition. Once you know what you've got, there are plenty of ways to improve it (see p27).

WHAT IS SOIL?

Pick up a handful of healthy soil and, basically, you've got a Walt Disney movie going on. It's a mixture of organic matter, minerals, water, gas, and thousands of insects, worms, and microorganisms. Soil is basically topsoil and subsoil (see right). While soil types vary depending on the geology of where you live, often your garden won't be simply one sort. Topsoil is what you need to be concerned with, and it can be divided into six main types. Two of these topsoil types, silt and peat, are rarely found in gardens, but you may have one or more of the other four (see opposite).

Topsoil is the layer of earth nearest the surface of the ground. It is rich in organic matter and full of life. The depth of topsoil can vary but is generally 2–5in (5–12cm) deep.

Subsoil is the layer found beneath topsoil. It is devoid of life and therefore not much use to gardeners.

ASSESS YOUR SOIL

The quickest way to assess the soil in your garden is to feel it in your hands and rely on your senses to tell you what kind of soil you have. Soil types vary around a garden so assess handfuls of it in different places to get the complete picture. This test couldn't be simpler to do, and it doesn't cost a penny.

2 Rub the soil between your index finger and thumb. Does it feel like clay? Or, does the soil feel sandy, or is it full of grit or stones?

1 Take a handful of soil from the surface, no deeper than about 4in (10cm). Feel the texture and look at the composition and color. Compare this handful of soil with the soil types described below.

3 Compact the soil in your fist to see how much moisture it contains. When you reopen your fingers, does the soil hold its shape, like clay, or slip through your fingers, like sand?

Sandy soil is low in nutrients and often on the acidic side (see p26). Light and easy to dig, sandy soil warms up quickly in summer but dries out easily.

Loam soil is a combination of clay and sand. It does not suffer from the extremes of the two types it is made up of and is very easy to work with.

Clay soil is rich in nutrients but is heavy and hard to dig. It retains water, draining slowly, and can take a long time to warm up in spring. In hot weather, clay can bake hard.

Chalky or lime-rich soils are very alkaline (see p26) and can be light or heavy. Often full of stones and nutrient-poor, this soil type drains freely and warms up quickly.

TEST YOUR SOIL

As well as finding out what type of topsoil you have (see p25), it's important to find out its pH (acid, neutral, or alkaline; see right). A basic DIY soil-testing kit from a garden center is all you need, and you can do the test any time of year. Soil type can vary around the garden, so testing it in a number of places gives a more accurate picture. For an average suburban garden, I'd recommend testing in three to six different places; the bigger the garden, the more areas you should test. Many plants will grow on either side of neutral, but the more extreme the pH, the more limited your plant range becomes. Once you know the pH levels in your soil, you can start researching which plants will do well in your soil type.

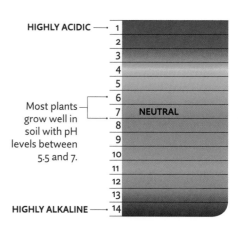

HIGHLY ACIDIC — 1 2 3 4 5 6 7 NEUTRAL 8 9 10 11 12 13 14 — HIGHLY ALKALINE

Most plants grow well in soil with pH levels between 5.5 and 7.

DESIGN · BUILD · ENJOY

YOU WILL NEED

- trowel
- plant pots or small plastic bags
- labels
- soil-testing kit

1 For each test, dig down to 4in (10cm), past any surface stuff that may interfere with the reading. Take a sample and put it in a bag or pot, labeling where it came from.

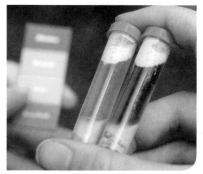

2 Follow the manufacturer's instructions on your particular kit for testing the samples and interpreting what the results indicate. Then keep a note of your findings.

GROW FOR YOUR SOIL

Some plants do better in acid conditions while others prefer a more alkaline soil. Healthy soil normally ranges from a pH of 5–8.5, and while it's possible to influence soil, I just wouldn't. It's always better to garden with what you've got. If a particular plant takes your fancy but doesn't like your soil, you can always grow it in a pot or raised bed filled with suitable soil. Garden centers sell ericaceous compost for acid-loving plants and multipurpose compost for all others.

ACIDIC

Rhododendrons need slightly acidic soil.

Camellias need slightly acidic to neutral soil.

IMPROVE YOUR SOIL

One of the most important things you can do for your soil is add organic matter. Add a layer of compost year after year and you will reap the benefits. Digging organic matter, such as homemade compost, well-rotted manure, leaf mold, or mushroom compost, into the soil boosts nutrient content, which means you will not have to spend money on soil fertilizer. As an added bonus, a good layer of organic matter can help retain moisture. If your soil test shows you have mostly clay or sandy soil (see p25), digging in organic matter will improve its structure, make it easier to work with, and encourage healthy plant growth. Worms and soil microorganisms absorb organic matter, and in time the weather can wash it away, so every 12 months you will need to add more. If your soil has poor drainage, working in lime-free grit can really help improve it.

WHAT ELSE TO LOOK FOR

- **Alien soil** often appears in new builds where topsoil has been stripped and replaced with something completely alien to the area. To find out whether your soil is alien, compare what's growing in your garden with a neighbor's, or local soil in the wild.

- **Compaction** by heavy machinery crushes soil, reducing space for water, oxygen, nutrients, and drainage. Digging soil over or aerating it with a tiller will open up the soil's structure again.

- **Areas of poor or no growth** may indicate the soil has a nutrient deficiency and needs an increase in magnesium, nitrogen, and/or potassium.

- **Patches of vigorous nettles** are a good sign. Their presence shows that the soil is rich in nitrogen, a nutrient that can easily be lost. Nitrogen in soil promotes green, leafy growth.

> *Soil type can vary around the garden, so testing it in a number of different places gives you a more accurate picture.*

ALKALINE →

Magnolias prefer slightly acidic to neutral soil.

Clematis flourishes in neutral soil.

Lilac prefer neutral to slightly alkaline soil.

Polemoniums prefer to grow in a slightly alkaline soil.

ASSESS YOUR GARDEN'S CLIMATE

Understanding the environment, prevailing winds, temperatures, and microclimates in your garden area cannot be done in an afternoon. It should take you at least a year, but your time and patience will help you shape your design and choose your plants.

Your garden's climatic conditions—and microclimates—not only determine which plants you can grow but will also influence your overall garden design. Sitting in the dampest, most rain-swept part of your garden area is never going to be a good idea, so by discovering more about the specific weather patterns in your garden, you can make sure you create comfortable places to spend time in.

THE WATER TABLE

The natural level in the ground where water sits is known as the water table. If you dug a deep pit, there's no doubt you'd eventually find it. However, for our purposes, the easiest way to understand the level of your water table is to be mindful of how plants behave. If your lawn stays bright green during a hot, dry summer, you probably have a high water table. If it wilts early in the season, this may indicate a low water table.

① MEASURE **AIR TEMPERATURE**

Some plants cope better with extremes than others, so variations in a garden's air temperature will affect the range of plants you can grow. To measure these variations, you'll need a thermometer that shows both minimum and maximum temperatures. Place it (or the sensor, if you're using a digital one remotely) outdoors out of direct sun. You can take a reading any time of day, but it's best to be consistent to keep track of how temperatures change over a 24-hour period during each season. What are the hottest and coldest readings? Are these usual for the time of year?

② DETECT **PREVAILING WINDS**

Wind direction affects the air temperature. When we get an east wind in Lincolnshire where I live, it is bitter. And when the wind blows mostly from one direction, it's called the prevailing wind. Note down the direction of any prevailing winds, how exposed your garden is to them, and whether any buildings or structures funnel wind anywhere. This information will have an impact on your design, as you may want to put up permeable screens and create cozy areas or grow plants that look lovely wafting in a breeze.

Knowing your garden area's climate helps you choose plants that will flourish. If your garden is damp and shady, plants such as this fern (*Dryopteris affinis*) will thrive.

HOW TO "READ" YOUR GARDEN

Ask yourself these questions to familiarize yourself with your garden's unique microclimate.

WHERE IS THE **SUN**?

Areas facing south get more light and warmth; areas facing north get less.

 ARE THERE ANY **THERMAL MASSES**?

Fences or brick walls facing south and west can trap and reflect heat, while north-facing walls offer cooler temperatures with less fluctuation.

 IS THE GARDEN **WINDY**?

The direction and force of the wind can damage plants and dry out soil, so in very exposed areas, you might need to think about introducing permeable barriers to shelter you and your plants.

 DOES **RAINWATER COLLECT** OR **DRAIN AWAY** EASILY?

Wetter areas take longer to warm up, so be aware of how easily or not water drains away. Areas that drain too freely may be susceptible to drought.

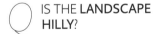 IS THE **LANDSCAPE HILLY**?

Even beyond your walls, hills or hollows can affect rainfall, while dips can collect cold air to form frost pockets.

③ GAUGE RAINFALL LEVELS

Rainfall can vary a lot from region to region, and understanding this will help you when it comes to planting and maintaining your garden. You can simply use a bucket to catch rain and a ruler to measure it, but a proper rain gauge (from most garden centers) is more accurate. Whenever rain has fallen, check the gauge (or bucket) at the same time of day. Understanding the rainfall in your garden year on year will help inform your plant choices.

④ DISCOVER RAIN SHADOWS

An area that rain is deflected away from is called a rain shadow. Wind, structures, and plantings can all deflect rain. Some areas may remain dry despite rainfall; others catch rain running off a hard surface and stay damp longer. Look for rain shadows after a downpour when damp spots are visible. Are there places in your garden with awkward spaces, buildings that deflect rain, or dense coverage (hedges, trees, or pergolas)?

> *By discovering more about your garden's specific climate, you will be able to create comfortable places to spend time in.*

WORKING WITH
CLIMATE

This city space was both sheltered and damp, so I used the shadows of the open, airy trees to play with light on the modern concrete surface and used water extensively to help reflect light. The plants were chosen not only for the damp conditions but also to bring light, movement, and contrast to the space.

The seating was designed as single units that could be used individually or brought together and could be moved around the space depending on whether shade or sun was wanted.

Limited color palettes can work well in smaller spaces as they rely on a lot of contrast. I chose plants that thrive naturally in damp, woodland conditions.

Polished concrete pads gave me a great reflective surface to bounce light around the space. A cantilevered step hovers over the pool, placed to make you pause and enjoy the water.

THINK ABOUT YOUR SETTING

The boundaries, views, and location of your garden area play an important part in how your design will look and feel. What goes on outside your garden is as important as what happens inside, so figure out what you can use to your advantage or what you might need to disguise.

Take a good look at the boundaries and your views, and consider the location in which your garden area sits. Think about how these factors will affect the design you're aiming for and the overall feel. Chances are that whatever you look out onto, good or bad, you'll be unable to change, so see this as an opportunity to get creative. There are many ways to take advantage of lovely outlooks, which is often called "borrowing a view," and ways to distract the eye from things you'd prefer weren't visible.

LOOK AT YOUR **BOUNDARIES**

Boundaries are used for security, privacy, screening, and defining the extent of a property and have a huge impact on the overall look. Interestingly, the smaller your garden area, the more important your boundaries become, since they are much closer and therefore a bigger part of your design. They can merge into the views beyond, creating a sense of flow from your garden to the landscape, or be used to create a sense of enclosure. For me, the most beautiful gardens are those whose boundaries "disappear," so the edges of the garden area merge or drift into the space beyond.

Maybe mix the materials in your boundary to add more focal interest, or lose your boundary by cloaking the wall with climbers.

USE (OR LOSE) **THE VIEWS**

How you use or lose a view can make or break your design, so take a good look around from within your garden as well as from inside the house to assess what is visible beyond your boundaries. There may be something out there you'd like to borrow, such as a tree or church. Or there might be some ghastly eyesore you would prefer not to see. If you plan to screen off such a view, think about whether you might be creating more shade. If you are using plants for screening, decide whether you want year-round coverage or just in the warmer months when you will often be outside.

A climbing wisteria in full bloom provides a colorful boundary that draws the eye in warmer months.

> *What goes on outside your garden is as important as what happens inside it.*

NEED TO KNOW

Boundaries affect other people around you so are subject to legal jurisdiction in the form of deeds and regulations. Before making any changes, think about the following:

○ **How secure your garden area needs to be and whether you need to keep pets or children safely enclosed.**

○ **Where your exact boundaries lie. There are legal implications around boundary lines. Check your deeds before annoying the neighbors.**

○ **How tall you would like to build your walls and what impact this will have on your neighbors.**

○ **Local authority regulations regarding height restrictions.**

○ **Shadows your boundaries may cast on your garden area and next door.**

EVOKING **TIME AND PLACE**

Finding out about your local area isn't essential to the design process, and some of you might prefer to ignore this next bit. However, I like to find out about any traditional building materials (such as brick, stone, slate, or flint), since these may be linked to the local geology and the practicalities of using what's on hand. For example, in an area that's mainly limestone, you may see a lot of great stonework. I also try to find out whether there are any local building methods linked with such materials, as this information really helps develop my designs. I hunt for architectural details as well. Walls, doorways, finials, gates, or windows are excellent places to spot such details.

Creating a design driven by local material or historical industries, and featuring an architectural detail or emblem helps your garden stand the test of time. Growing plants endemic to the area is another way to harmonize a garden with its locality.

Ask ADAM

There's so much to think about when you're mapping out your garden area. You need to look at the space available, test the soil, and find out about the garden's climate and microclimates, as well as researching boundary title deeds.

DESIGN · BUILD · ENJOY

WHAT IF MY SOIL IS LIKE RUBBLE?

New-build houses often have gardens with lifeless "alien" topsoil, but any type of soil, no matter how poor, can be improved by digging in organic matter, such as well-rotted manure, leaf mold, mushroom compost, or homemade compost. To help plants take up nutrients, you can also add mycorrhizal fungi when you plant them.

DO I REALLY HAVE TO DRAW UP A SCALE PLAN?

It may seem a little over the top to draw up a proper scale plan of your garden, but it's a key part of the design process. Your plan will not only help you organize your space as effectively as possible, it will also ensure that your calculations are more accurate. Ultimately, this will save you money when buying in materials or getting quotes from builders.

If you can't afford the exact same materials used historically in your area,

use materials of similar colors

to help connect your garden to its location.

WHAT IF I CAN'T DO IT ALL IN ONE GO?

Don't worry if you can't do everything like measuring the garden, soil testing, and checking who owns the boundaries all at the same time. It's very unlikely any of this will change. It's better to take your time getting to know your garden area by sitting in various places and looking at it over a period of time so you can really understand how the sun and shadows move through the space. You'll also discover where your favorite spots are at different times of day.

> *It may seem a little **over the top** to draw up a proper scale plan of your garden, but it's a **key part** of the design process.*

WHAT ABOUT TREES AND SUBSIDENCE?

Most trees near buildings don't cause damage. Issues tend to arise in areas of heavy clay that have undergone prolonged drought, as tree roots seeking water dry out the soil underneath foundations. Buildings on chalk or sand rarely suffer tree-root damage, and if your drains are watertight, roots will generally not cause any trouble. Look out for new cracks around windows and doors. If in doubt, seek professional advice. When choosing a tree, think about how much space the roots need. I never plant a tree closer than 10ft (3m) to a home, and even then, I would place only a small tree near a property.

QUICK FIX

IMPROVE **YOUR** SPACE IN THE SHORT TERM

I always recommend taking 12 months to really get to know your garden area and to observe how it changes through the seasons. But if you're eager to get started, there are a few quick fixes. Have a general cleanup and remove as much clutter as possible. Clean stone or brickwork and fix anything that's loose. Use an edging tool to spruce up lawns. Sow annuals in pots and bare soil to bring instant color and to help keep weeds down. Prune any dead or diseased material.

KEY *knowledge*

○ Your soil is the single most important thing in your garden area, so make sure you understand what you've got and what you need to do to improve it and keep it healthy.

○ The orientation of the sun is the best guide to the nicest places to sit in the garden. Areas facing south get more light and warmth; areas facing north are colder and darker.

○ My favorite gardens are those that appear to have "lost" their boundaries. To make them disappear, cover boundaries with climbers or climbing shrubs.

○ Research the history of your local area and use materials, emblems, or colors that echo the local architecture and geology.

DESIGN FOR PEOPLE

> " *Ask yourself who will use the garden and how they'll want to use it.* "

INTRODUCTION

In order to create a garden area that people are really going to enjoy, you need to think about who will use the space, what they will use it for, and how much time and money you can spend on it. There is no point designing a garden area that doesn't fit with you or your family's needs.

I often think that most of us lead two lives: the one in our heads and the one we lead in reality. As you figure out who you're creating the garden area for, you have to be absolutely honest with yourself. Your design might look lovely on paper, but if it doesn't meet the needs of you and your family, it isn't going to work in the long run.

When designing a garden area for a client, I ask countless questions as I try to build a picture of their lives, wants, and needs. That's what I want you to do, too. Interview yourself: ask who will use your garden and how they will want to use it. Even if some of the answers may seem obvious, they'll help your plans become more specific and tailored to your needs. By asking yourself what you want from your garden area and writing your answers down, you'll start to pinpoint what your garden specifically needs to include to allow you and your family to do the things you enjoy.

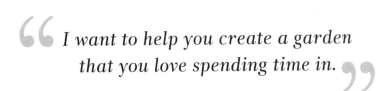

I want to help you create a garden that you love spending time in.

WHO IS YOUR GARDEN AREA FOR?

For me, the best gardens are well loved, well used, and uncomplicated. Garden areas can draw people out of the house because they make them feel comfortable, relaxed, and happy. Take a moment to think about who will be using the space and how best to meet their needs.

We are used to putting together rooms in our homes that answer people's needs, such as a well-equipped kitchen for cooking and eating in or a sitting room with comfortable seats, TV, and a sound system. So think about your outdoor space along the same lines.

To do this, go back to basics. Who will be using the garden area? How old are they? Think carefully about them and their lifestyles. As you work through the questions, you'll start to build a picture of all the wants and needs your garden has to fulfil.

Q ARE YOU SINGLE, A COUPLE, OR A FAMILY?

The number of people your garden area needs to accommodate will affect how you arrange it and what space, if any, you include for recreation areas or lawn. It will also affect the size of your seating and dining areas.

Think about

○ how many people will use the garden area on a day-to-day basis

○ how you and your loved ones prefer to spend time together

Bear in mind

If you are likely to all hang out together as a family, you will need a communal area where both adults and children feel comfortable, though it might be more realistic to provide a separate space for children to play away from the grown-ups. And don't forget your own needs. If you are creating a garden area for a large family, you might want to consider creating a quiet, peaceful space where you can escape for a moment and spend time alone.

A communal area can really bring people together in the garden.

Q ARE YOU A GARDENER?

Your garden area needs to match your ability—and your enthusiasm—for gardening, along with the time and budget you have available. As well as influencing your planting choices, this question also has an impact on whether you want to include features such as a lawn, vegetable beds, trained fruit trees, or a greenhouse.

If you spend all your spare time in the garden area, design accordingly; but if you are really busy, be realistic and keep your garden low-maintenance.

Think about
○ how keen a gardener you are
○ how much time you have

Bear in mind
Be honest here. There's no point in starting out by creating a high-maintenance garden if you don't have much experience. It's best to begin with plants that perform really well but are easy to maintain. Then, if you like getting more stuck in, gradually build up your knowledge and develop your skills so you can take on things that need more attention and know-how, such as propagating new plants or pruning fruit trees.

Q WHAT'S THE AGE RANGE?

Consider the age range of the people who will use the garden area, including children, teenagers, parents, and grandparents. How will their age influence the way they use the space?

Think about
○ the ages of the main users of the garden area
○ any children or grandchildren; how old are they?
○ any family or friends who are quite elderly

Bear in mind
While some younger family members will happily spend hours playing outdoors, others may need some encouragement to draw them out of their bedrooms and into the fresh air.

Involve your family in the garden to draw them into the space and connect them to it.

Think about providing a space that is just for them, such as a den or tree house for younger children, or a comfy area with beanbags or hammocks for teenagers.

" While some younger family members will happily spend hours playing outdoors, others may need encouragement to draw them outside. "

Q HOW **PHYSICALLY ACTIVE** ARE YOU?

If you or a relative have reduced mobility, width of paths, surface types, and use of steps should all be informed by your accessibility needs. You want to create a garden area that you find easy to care for and that you will be able to fully enjoy.

Think about

○ **any wheelchair use or physical challenges to factor in**

○ **how difficult you find it to walk, kneel down, or bend**

Bear in mind

Accessibility isn't just about avoiding trip hazards and including handrails on steps or slopes, although that is a good start. If you or a relative find it difficult to bend, kneel, or walk, you might want to create raised beds, which are easier to plant, maintain, and view up close from a seated position. If there are wheelchair users, paths need to be wide enough for them to move around. The surface of paths is a consideration, too. Wheelbarrows, wheelchairs, and strollers are easier to maneuver and move around on paved surfaces than they are on gravel.

Raised beds offer relatively easy access and support, as they can be reached without too much bending or stretching.

> *Accessibility isn't just about avoiding trip hazards and including handrails, although that is a good start.*

Q WHAT ABOUT YOUR **HOBBIES?**

You're designing a garden to spend time in, so why not design it to suit the interests you already enjoy? This question may help you decide whether a single multipurpose open area, such as a lawn or patio, might suit your family better than smaller spaces designed for specific activities.

Think about

○ space for exercise

○ family sports and games

○ any wildlife lovers in the family

Bear in mind

The obvious choice when designing for sports and exercise might be to include a large lawn area, but earmarking a quiet, private space for gentle exercise, such as yoga or Pilates, could be just the encouragement you need to get up and work out each morning. Or if you love wildlife, why not plant wildflowers in a designated nature corner to attract birds, butterflies, and bees?

Q WHAT ABOUT YOUR **PETS?**

Pets are a part of the family, and while you may not want to design the garden around them, it's worth considering the role they play in your outdoor space.

Think about

○ the size of your pets

○ whether they live entirely outdoors, or roam freely between house and garden

○ any mess they're likely to create when let loose in the garden

Bear in mind

Remember that all pets need plenty of space. If you have a rabbit, for instance, consider where the hutch will stand, and allow space on the lawn for a rabbit run. Don't forget about the garden area boundary when planning a pet-friendly garden area; walls and fences need to be secure to prevent escape.

Balance the needs of your pets with those of your ambitions as a gardener. Don't expect them to change their habits just because you've changed the garden.

DESIGNING FOR
PEOPLE

I designed this garden to be a place where a family can enjoy the outside world, from cooking on an open fire to clambering around in the rock pool. The undercover seating and fireplace make it a comfortable place that can be enjoyed all year round.

This open-sided stone shelter, with a fireplace and a cantilevered oak roof, is the ideal spot for friends and family to relax and have fun together.

I selected this stone to evoke family memories of a special place and then used it in both raw and treated states throughout the design, which brought a wonderful sense of place.

The naturalistic planting and the green roof combined to fulfill the ambition to create a diverse habitat for local wildlife.

WHAT IS YOUR GARDEN AREA FOR?

Forget the practicalities for a moment. What do you want from your garden area? How would you ideally like to use it? Keep an open mind and be honest about the garden area you wish you had. Then start thinking about how you might be able to achieve it.

So many of us begin creating a garden area without really considering its function in detail. I try to get people firstly just to think about the garden area as a space. This may sound strange, but I find that the moment we think "garden," our minds go to the practicalities of sheds, compost heaps, storage, and so on, which—although very important—don't tell you how you want to use the space.

If instead you think about how people will use the space and how it can fulfill their needs, it will open up more possibilities for your design.

Q WHAT DO YOU **WANT FROM THE SPACE?**

Take the time to really think through your answer to this question. Try to make sure that everyone who's going to use the garden area is involved, and don't worry if the wish list gets ridiculously big. You can always whittle it down later as you begin taking into account considerations like time and budget.

Think about

- whether you want a place where you can relax
- whether you want to use the garden area for entertaining
- whether you want to grow lots of flowers
- whether you want to grow fruit, vegetables, and herbs
- whether you want a play area
- whether you want to attract wildlife

When you have written up your list of ideas, begin to dig a little deeper into all the options to figure out how each wish could potentially be achieved.

Maybe flowers are a priority on your wish list, or are you more interested in growing your own fruit and vegetables?

Q IS YOUR GARDEN AREA FOR PLAY?

A dedicated play area can be a great feature to draw kids outdoors, but there's no point trying to design a play area that, in reality, won't actually fit in the space you have. How will your children actually use your garden area? Think about the games they would play and the activities they would enjoy and make that the starting point for your plans.

Think about

- how many children will be playing there (yours and their friends)
- allowing space for fixed structures, such as a set of swings, a trampoline, or a jungle gym
- leaving space for ball games
- where you could put a wading pool in summer

Bear in mind

Over the years, our family has changed the games we have played to suit the size of our garden area, instead of trying to shoehorn an entire soccer field into the wrong-sized space. If you don't have room for a dedicated play area, get creative or go upward by using existing structures to hang up swings. For example, if you have a mature tree, you could build a tree house.

To make a dining area feel more intimate and private, you may want to add an overhead structure to enclose the space.

Q IS YOUR GARDEN AREA FOR ENTERTAINING?

To entertain guests—be it lazy Sunday lunches or evening drinks under the stars—you need to choose the right spot in your garden area and allow plenty of space so your friends and family can comfortably move around and enjoy themselves.

Think about

- how often you would like to entertain
- how many guests you would expect to invite on average
- whether you would prefer all your guests to eat together at a table or sit around lounge-style
- whether you want to cook outside
- what times of day you like to entertain and what times of the year
- whether you need to be able to leave seating out all year round or if you would prefer to store it over winter

Bear in mind

When it comes to planning an area for entertaining, I tend to take the average number of people we regularly have over and use that to determine the amount of space I need. And don't make the mistake of allowing just enough space for the table with the chairs pushed in. How many times have you sat on a patio, moved your chair back from the table, and nearly fallen into the flower bed? Make sure you allow plenty of space and, if in doubt, always add more. As for the time of day or year, don't forget the practical aspects. You may need to factor in some shade for midday sun or a fire pit if you don't want your guests shivering on a cool fall evening.

When designing your wildlife garden, consider which flowers to plant to attract bees and butterflies and where you can hang bird feeders and put bug houses safely.

Q IS IT **ALL ABOUT** THE **WILDLIFE**?

Songbirds, butterflies, bees, and croaking frogs bring a garden area alive and are a great way to educate people about the importance of the natural world and how gardens can act as wildlife corridors. There are lots of ways to make your garden area wildlife friendly, so choose the features that work best for the space.

Think about

○ **how to accommodate your cats but still see birds feeding in your garden**

○ **whether your children are old enough to be around an open pond or water feature**

○ **where you could put bug houses and bird feeders so you can see them from your house**

○ **planting trees and bushes for winter berries to feed the birds**

○ **whether your fences are hedgehog friendly, with openings for them to pass through**

○ **how you can get the best from the space in terms of planting**

Bear in mind

If you have cats, make sure you position bird feeders high off the ground where cats cannot reach them. Locating bird feeders in prickly bushes, such as holly, will also deter cats. You can put obstructions on the supporting pole of a bird table, called "baffles," which will prevent cats and squirrels from climbing up. If you have young children, an open body of water such as a pond could be dangerous; a high-sided water feature may be a better option.

Q IS GROWING **FRUIT AND VEGETABLES** A PRIORITY?

There's nothing that quite beats the taste of freshly picked produce, nor the sense of achievement you get from cultivating your own personal harvest. If you're thinking of incorporating a kitchen garden into your outdoor space, take a moment to consider what you want to grow and how much you actually need to grow.

Think about

○ what fresh produce you like to eat and what fresh herbs you cook with

○ whether you have space for climbers, such as peas or scarlet runner beans

○ whether you want to grow different crops each year or devote a permanent space to perennial vegetables, such as asparagus, rhubarb, or fruit bushes

○ whether you want to interplant crops among perennials in your borders

Bear in mind

Think carefully about how much space you can realistically dedicate to growing crops. Don't think you have to keep your flower garden and vegetable garden separate, either. If you're short on space (or if you just want to get a little creative with your planting), try combining your crops with your ornamentals: scarlet runner beans among your climbers, kale mingled with marigolds, and so on. As for herbs, a great introduction to the grow-your-own lifestyle, consider planting these close to the house so you can easily step out and pick just as much as you need.

A dedicated area for growing vegetables is a must for some, but you can always plant your edibles in with your ornamentals if you want to mix it up.

Q IS THERE **ROOM TO GROW?**

It's easy to plan for the big items on your wish list: areas for entertaining and play; planting ideas for wildlife; and kitchen gardening. But it's those smaller moments, the fun times spent together in the garden as a couple or a family that really make it special. Put down your pencil and imagine everything that could happen in your garden over the years. Then start to plan a space that could lend itself to those future memory-making moments.

Think about

○ whether you have allowed enough room for seasonal activities, such as setting up a summer wading pool

○ including a barbecue

○ whether there is an area for a firepit you can all sit around

○ whether there is space for family games, such as table tennis or badminton

Bear in mind

What matters here is space and ensuring there is enough room in the garden area to allow for those impromptu family moments. Your interests and hobbies may change from year to year, but an open space, such as a lawn area, can become whatever your family needs it to be as you all grow.

HOW MUCH TIME DO YOU HAVE?

Once you have your wish list, it's time to be realistic. Although you might want an outdoor wonderland for everyone to enjoy, you need to be honest about how much time you can spend creating your garden area, and how much time you can dedicate to maintaining it.

Gardening is all about time, from the passage of weeks and months as a plant comes into bloom to the decades of patience required to watch a tree grow to full maturity. Garden design and planting are also ongoing processes. Things don't stop the moment you plant your last bulb; your garden area will evolve as you and your family's needs change. Remember, too, that once you have your design, you can break down the build and do it one bit at a time. As you care for your finished garden, don't be afraid to make adjustments to suit your changing needs.

> *Garden design and planting are an ongoing process— it doesn't stop the moment you plant your last bulb.*

Q HOW **QUICKLY** DO YOU WANT TO **CREATE IT?**

You may want to build your garden as soon as possible, or it may suit you better to break it down into various chunks of work.

Think about

- whether you are eager to have the garden design finished as soon as possible or whether you can take your time and follow a stage-by-stage approach
- whether you need to bring building materials through the house
- whether there is a special event that means the garden area must be finished by a certain date
- which areas you need to prioritize, and which can wait, if you can build the garden bit by bit

Bear in mind

It is important to have a master plan, but the actual build can be broken down into projects. Breaking down elements of the garden area into stages can allow you to budget over a longer period and save up money. However, bear in mind that buying building materials in bulk is always cheaper, and if you will need to bring stuff through the house, you may not want to have to clean up repeatedly. If you are planning to redecorate the house, it may be better to finish the mucky business of garden building first.

Q HOW **BUSY** ARE YOU?

How much time do you actually have to look after your space once you've created it? Think about your commitments and figure out how and when you will be able to care for your garden.

Think about

- how often you are away on the weekends
- whether you have to do a lot of running around and being a "taxi" for your kids
- how many vacations you take every year
- what time of year you tend to go away
- whether you already enjoy puttering in the garden for hours
- how long your current "to-do" list is (not just garden jobs, but all the tasks that you know need doing—do they mount up quickly?)

Bear in mind

If you are already hands-on in the garden and have plenty of time to dedicate to its upkeep, then by all means go ahead and create a high-maintenance space you'll enjoy caring for. But if you know, deep down, that you won't have that much time to spare, don't worry. It's perfectly possible to create a beautiful garden that places fewer demands on your time, with a higher ratio of hard landscaping to planting areas, that needs only basic watering, feeding, and weeding to keep it in shape.

Q HOW **LONG** WILL YOU BE THERE?

The amount of time and money you spend on your garden should reflect how long you plan to use it. It may seem obvious, but you don't want to spend thousands of dollars and hundreds of hours creating and maintaining a garden that you'll say goodbye to in a few years.

Think about

- how long you intend to stay at your property
- whether all of the people who are currently using the garden area will continue to live there
- whether any children or teenagers will eventually go off to study or leave home

Bear in mind

If you intend to stay in your property for only a few years, you may not want to spend as much money on the garden as you would if it were your "forever home." You may want to think about elements—outdoor furniture, potted plants, and so on—you could move from one garden to another. Even if you do intend to stay for the long term, you may not necessarily want the garden to look the same in 10 years. Think about how it could potentially evolve with you and your family, so you don't need to redesign it over and over again in the future.

Put plants in containers if you think you might want to take them with you when you move.

Ask ADAM

Making sure the garden area works for the people who use it is one of the biggest challenges when designing a garden. It's not just about meeting practical needs but also about making sure the space is a joy and doesn't become a chore or a drain on your finances.

If you sit down as a family and get everyone to discuss **what they want from the garden area,** they are much more likely to use it.

HOW DO I GET MY **FAMILY TO AGREE** ON A DESIGN?

Sometimes it's difficult getting all family members to agree on what the garden area is for, how it should be laid out, and what kind of style it should have. Someone once said that in marriage you can either be right or be happy: I think this is true of garden design, too. The only way to keep as many people as happy as possible is to make compromises. Find the main things that can be agreed upon. When you make your lists, see where the points of crossover are and focus on those instead of getting hung up on specific details.

WHAT IF I HAVE A **TINY BUDGET?**

Good design doesn't have to cost a fortune. The main thing is to have a space you enjoy using. Whether you have expensive stone slabs, salvaged concrete ones, or even gravel, it doesn't really matter. If you design your garden area well, then you can always build it gradually in stages to allow you to take the time to save up money for things you really need and want to have.

HOW DO I GET MY **KIDS OUTDOORS?**

One of the best ways to get kids outdoors is to provide them with some sort of a den. You can buy playhouses, but anything from a homemade tree house to a tepee made of sticks will do. Another way is to get them growing vegetables or cooking outdoors. I love making pizza with my kids and have put a pizza oven in my garden area so now all their friends come around, too.

> *Good design doesn't have to **cost a fortune**. The main thing is to have a space **you enjoy using**.*

QUICK FIX

HOW TO FALL IN LOVE WITH **YOUR** GARDEN

When designing your garden area, make sure it's a space you really want to spend time in. Ideally, your garden area will become a place that you can't help but be drawn into whenever you look out the window. If there's a comfortable place to sit, ideally in a warm secluded spot, it will mean there's always a destination. Food eaten outdoors is always tastier, so make sure you have a decent table and enough seating for everyone. If you can cook outside, even better.

 **HOW WILL I FIND TIME TO DO THE GARDENING?**

A lot of us have really busy lives, and people who are not used to gardening often worry that they won't have enough time for it. However, you can create a garden that needs as much or as little time as you want to give it. Some plants need a lot of attention while others can simply be left to do their own thing. Just choose your plants carefully so you don't inadvertently make a rod for your own back. And hard landscaping, such as paving and bricks, needs cleaning only a couple of times a year to keep it in good condition.

KEY *knowledge*

○ I think most of us live double lives: the one we dream of and our everyday life. Be as realistic as you can about how much time you have to spend in the garden.

○ Think carefully about all the people who will use the garden area and for how long.

○ Make sure the garden area is designed in such a way that it doesn't become a chore. If you're not a keen gardener, then avoid making a lot of work for yourself because you'll only end up resenting it.

○ A great garden that is well used and well loved doesn't have to cost a fortune. If you plan it well, you can use your resources to best advantage and the garden area can be built in stages or upgraded in the future with better materials.

DESIGN

FIND YOUR
STYLE

> *Style ideas can come from anywhere: magazines, websites, even a tile pattern on a kitchen floor.*

INTRODUCTION

Just like a room inside a home, a garden is made up of various elements that, when combined together, give it a sense of place and style. It's easy to buy things in an ad hoc way, but for a more cohesive look, you'll need to think through what you'd like to include and the overall effect they'll create.

Every element in your garden area—paths, boundaries, steps, water features, structures, lighting, and furnishings—will play its part in how the space feels, so you should take your time when deciding what you want to include and where you want to position it.

In this section, I'll share with you a selection of galleries that show how versatile you can be with your chosen elements and how each one can help contribute to an overall style. After that,

I'll show you how to create a mood board (see pp82–83), an essential part of the design process. I tend to use them to fine-tune my choices and decide how the colors, textures, and materials of my chosen elements will work together. This is your chance to get creative and really play with different ideas, though I would suggest that, once you start to hone in on an individual style, try to keep to a limited palette of materials. It always pays to keep things simple.

" *Look for materials and details in the surrounding landscape that will ground the garden and give it a timeless feel.* "

FINDING THE RIGHT MATERIALS

Before you choose which hard landscaping materials you want to use in your design, it helps to know a little about them. While aesthetics are important, you'll also need to keep in mind practicalities such as cost, durability, and how easy a material is to maintain.

> *If you use classic patterns of materials, such as stone, clay, wood, and even some types of metals, then your garden is less likely to date.*

HARDWOOD

Wood taken from broad-leaved trees, including oak, chestnut, and ash, is known as hardwood. This type of wood typically feels high quality; it's not surprising to discover, then, that hardwood is more expensive than softwood (see right). Smooth (planed) wood can work well with a more modern look, while rough (sawn) wood lends itself to a more rustic aesthetic.

Since wood is a natural material, it changes with exposure to the elements. Most hardwood "silvers" beautifully as it ages, but if you'd prefer to keep its original color, then you will have to treat it.

SOFTWOOD

The wood of coniferous trees—such as pine, fir, and cedar—is softwood and tends to be yellow or reddish in color. These trees grow faster and so are generally much cheaper than hardwoods. Like hardwood (see left), softwood is available both planed and sawn, depending on the look you want to achieve.

If you want to bring some color into your garden, softwood can look great, especially if you paint or stain it or work in some nice design detail. Most softwood will need to be treated if used outside—check with your supplier once you've chosen. It is normally pressure treated.

BRICKS

You can buy bricks in a wide range of colors, finishes, and sizes, so it's pretty easy to find something to fit in with your design aesthetic, from traditional to contemporary. Generally, you get what you pay for in terms of quality. It depends on whether you like the uneven texture of a handmade brick or the consistency of a factory-made one. If you'd like your garden area to feature brickwork that matches your house, be mindful that if you want to use it on the ground, it will have to be frostproof.

STONE

Depending on your budget and preference, you can buy different types of stone slabs including granite, limestone, and sandstone. The finished look of your garden area will depend not only on the type of stone you choose but also how it's cut. A natural riven finish fits a more rustic or traditional aesthetic. For a more modern look, try crisply cut stone, but take care where you position it—certain stone can become damp and slippery on a north-facing terrace. That said, mixing textures can look great in any style of garden.

CONCRETE

Paving slabs made of precast concrete come in many colors and finishes—everything from wood effect to sleek and smooth. Don't be put off by the word "concrete"; there are some fantastic effects and finishes available.

In addition to the precast products, you can also pour and polish concrete on-site for a crisp, contemporary look and to create more fluid shapes. For this, however, you might want to call in the specialists.

LOOSE GRAVEL

Gravel is a cost-effective way of covering large areas and comes in various grades. You can walk more easily on smaller-sized gravel, making it better for high-traffic paths. Chunkier gravel, meanwhile, might be useful if you want to encourage people to slow down and enjoy their surroundings.

Look for gravel from your local area, or with a blend of colors, as this works with a broader range of materials. Buying gravel in bulk always works out cheaper.

SELF-BINDING GRAVEL

Self-binding gravel is a mix of large, small, and fine particles that form a solid surface when compacted. Although it's more expensive than loose gravel, it creates a hard-wearing surface that weathers well, so when compared with paving slabs, it can work out to be relatively cheap.

If you use this gravel near the house, make sure to place paving at door entrances so you don't bring particles into the house.

CORTEN STEEL

This remarkable material has a stable, rusty surface with a lovely, earthy tone. It can be made into all sorts of things—containers, screens, edging, and even water features. There are many ready-made Corten steel products, but it's also not as expensive as you might think to commission bespoke pieces from a metalworker.

One word of warning: Corten steel can cause staining, particularly if used in conjunction with wood or over paving.

TERRACES

Terraces and paving are usually close to the house and so should provide a stable, easily cleaned surface where there's usually heavy traffic. Make sure the style and materials are in keeping with the rest of the setting, and assess how the colors and textures will be affected by sunlight, shade, and rain.

" If you've got a brick house or wall, think about incorporating matching brick in the terrace edging. "

3

4

1 Reclaimed stone not only looks elegant but also gives a terrace a timeless feel.

2 To create a cohesive look, this terrace picks up on the brickwork of the boundary wall and house.

3 Poured concrete slabs give this staggered terrace a contemporary feel and creates a smooth, light-reflecting surface.

4 A coursed paving pattern can make a terrace feel wider, while light-colored slabs can really lift a dark space.

DESIGNING TERRACES

The terrace marks the transition between the house and garden area, and making a strong link to the architecture of the house should be one of your design considerations. It's also important to take into account the practicalities of size, shape, and materials, and aesthetics.

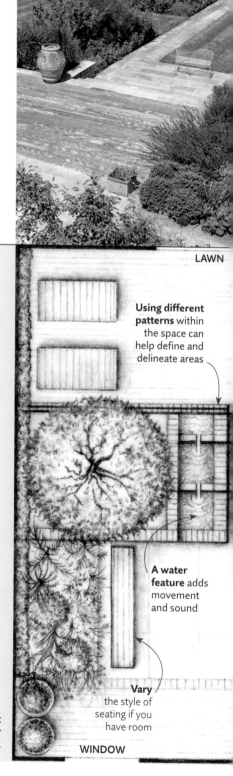

WHERE TO START?

First, figure out what you plan to use the terrace for and what size it needs to be. The direction your garden area faces can also affect the dimensions of your terrace. If it faces north and you have space, you might want to extend the terrace out of the house's shade and into the sun.

Your choice of materials is key, so take some time to experiment and consider your options from both a practical and aesthetic standpoint.

WHAT SHAPE?

Don't worry about creating fancy shapes. For me, strong, simple shapes always work best. And if you are building it yourself, you'll appreciate regular shapes that mean you don't have to do lots of complicated cutting of pavers.

WHAT SIZE?

As you start to design on plan, you'll need to think about what size you want your terrace to be. Roughly mark out the area to make sure it is in proportion to the house. Does the terrace look too small and lost? Or does its size overwhelm the house? Make sure it is the right size for its purpose, too. A common mistake is not allowing enough space for the dining area; always make sure you have plenty of room to pull chairs out from the table.

This sketch shows the different features you need to consider when designing your terrace.

LAWN

Using different patterns within the space can help define and delineate areas

A water feature adds movement and sound

Vary the style of seating if you have room

WINDOW

LET IT FLOW

Think practically about how you will move around the space. It's convenient to have a dining area on the terrace near the kitchen, but you don't want it getting in the way when you're going to other parts of the garden. Also check the terrace's position and orientation from inside the house. Be careful that the furniture won't block views of the garden area from the house.

PLAY WITH PATTERNS

If you lay coursed paving slabs across your garden area, it can make it appear wider; if you lay them going away from your house, it can emphasize the length. Think about the unit size of your paving. For a clean and contemporary look, choose large pavers or poured concrete so you have minimal lines. If you want a more traditional look, go for something more classic such as brick paving or granite setts.

WHICH MATERIALS?

Take inspiration from materials, colors, and patterns used on or in the house. If you have a blue slate roof, for instance, you may want to edge the terrace in blue bricks; or there may be a pattern on the kitchen flooring that you can repeat on the terrace. Perhaps choose materials traditional to your area, even if you put a contemporary spin on them. All these details help give the terrace a stronger connection to its setting. Always get samples of your chosen materials to see how they look in context, too. Bear in mind that a small sample of stone may not represent exactly what you'll end up getting, as natural materials can vary from batch to batch.

WHAT'S PRACTICAL?

Practical issues are very important, both in terms of color and texture, so make sure you understand how materials will work in a particular spot. If the terrace faces south, pale-colored materials can get blindingly bright in summer. If it faces north and you live in a region with high rainfall, the area will be shaded by the house and often damp, so avoid smooth stone that can become slippery.

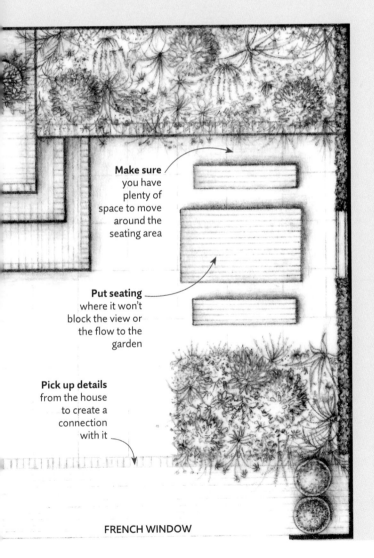

Make sure you have plenty of space to move around the seating area

Put seating where it won't block the view or the flow to the garden

Pick up details from the house to create a connection with it

FRENCH WINDOW

PATHS AND WALKWAYS

The design, materials, and shape of your paths all directly impact your garden area, visually and practically. In addition to being a way of getting from point A to B, a path can help divide the area, enhance the drama and geometry of a scene, gently wend its way to far-flung corners, or even create a focal pull.

1 **The coursed pattern** of the smooth paving draws the eye along the path, while the irregular edges are softened by the planting and pebbles.

2 **Using small bricks** offers an opportunity for tighter corners, which can help create a sense of movement and intrigue.

3 **The self-binding gravel** of this secondary path contrasts well with the smooth texture of the main stone walkway, while its pale color bounces light around the garden.

4 **Stepping stones** can turn a path into an adventure as you explore and enjoy the plants around you. They are best used for low-footfall walkways.

5 **Natural wooden planks** laid across the direction of travel visually exaggerate the width, creating an illusion of space. The warm tones of the wood set off the planting to great effect.

> *Give people a path to follow and their heads will come up to look at the plants instead of down to look at their feet.*

DESIGNING PATHS AND WALKWAYS

The most important aspect when designing a path is its purpose. Think about the path's destination, how quickly you want people to get there, and how heavily used the path will be.

WHY HAVE A **PATH**?

Paths determine the way people navigate a space. They are also a way to direct people around your garden area, steer them to focal points or views, or slow them down to enjoy a feature of the space.

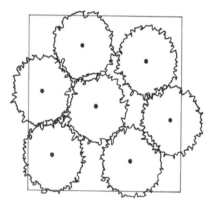

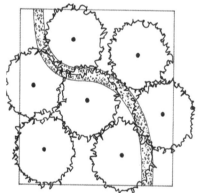

Without a path, people move haphazardly through a space. Think about how you'd walk through a wild area such as a woodland—most likely you'd be looking down for trip hazards rather than enjoying the views.

Adding a path gives a sense of movement and destination. As soon as you add a path, you don't have to think about where to put your feet and your head can come up instead, allowing you to look around.

WHERE DOES IT LEAD YOU?

Are you just going from point A to B? Will the path be used every day or only occasionally? A primary path should be roughly 3ft (1m) wide so that it is practical to walk along, although paths used less often can be narrower. Primary paths that are heavily used will need a hard-wearing material. Tie materials in with the overall look of the house and garden, and use focal points to invite people along a path.

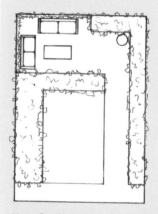

Use a focal point to draw people along a path and pull them into a space.

EDGING

You'll need to decide what kind of edging to use where a path meets other features, such as a lawn or border. A gravel path, for instance, may need edging to prevent gravel from spilling into beds. A lawn will need flush edging (no higher than the grass) to make mowing easier.

- Simple brick edging gives a classic, timeless look. You can create different effects by the way you lay the bricks.

- Wood edging is cheap and cheerful and can be used straight or curved.

- Steel edging is great for giving a smart edge to a lawn, but it needs to be installed carefully, especially if it is to go around bends and sharp curves.

- Small units such as granite setts are hard wearing, easy to use, and come in various shades of gray. They are particularly useful when creating tight curves.

- Stone edging is a good way to tie in with, for example, the stone used on a terrace and helps give a continuity of materials around the garden.

- Clay tiles can look great and come in lots of different shapes and sizes.

STRAIGHT OR CURVED?

While the quickest route from point A to B is a straight line, it might divide the garden area in a way that is not useful to your design. Curved paths can slow people down and bring a sense of movement into a garden. A focal point can slow them down midway along a path, while arches or a pergola can provide interest and a sense of mystery. Edge your path with planting to stop people from cutting the corner.

PAUSE FOR THOUGHT

Adding seating along a path can encourage people to pause as they move through the space, as well as invite them to enjoy different views and get more out of the garden. Irregular paving with planting to the side can soften the lines of a path. Using different materials and patterns can also create places to pause. If you're adding seating, remember to provide a good view or focal point to look at.

DIVERSIONARY TACTICS

Paths that are used only occasionally should still be part of your overall design, so plan them carefully. They lead the eye, invite possibilities, and tap into the power of suggestion. Materials can be varied, or match the primary path to link an area neatly as part of the design. Stepping stones might not be heavily used, for example, but they still provide a real focal pull, and on a practical level, they can help prevent wear and tear on your lawn.

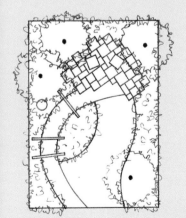

A **curved path** edged by planting stops people from cutting the corner, while arches can draw them through a space.

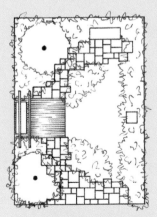

A **pause point** on a path can be delineated by different paving and an arbor, with a seat and a view.

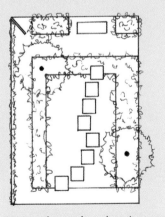

A **secondary path** can have its own character, distinct from the primary, practical path.

STEPS AND LEVEL CHANGES

There are lots of different ways to construct steps and deal with slopes in your garden. It's important to figure out exactly what levels you're dealing with and how steep they are. Whatever approach you choose, make sure it's comfortable to walk on and safe to use.

> *To save money when constructing level changes, think how you can use the material you've excavated from one place and relocate it in another (known as 'cut and fill') rather than paying to take it off-site.*

1 Wooden sleepers are a cost-effective way of making steps. They work best in a natural setting away from the house.

2 Natural locally sourced materials can create a traditional feel and give a sense of place.

3 Adding detail and pattern in brickwork can transform a simple set of steps into a focal point.

4 The boulders and informal planting surrounding these low, wide steps ties in this seating area with its surroundings.

5 By making the steps broader than the front door, this entrance feels more open and welcoming and also provides space for planted containers.

BOUNDARIES

Your boundaries form the backdrop to your whole garden, and your choice of colors, materials, texture, and patterns will affect the atmosphere of the entire space. Think about the purpose of your boundary, the effect its height will have, and how much money you want to spend.

> *Just like the walls of a room, your choice of boundaries will have a big impact on the look and atmosphere of your garden.*

1 Boundaries can be a feature in your garden, almost like a piece of sculpture. This see-through screen divides the space without making the garden feel smaller.

2 Horizontal fencing panels not only provide the perfect setting for plants but also make a garden feel larger.

3 Trained fruit trees can look striking against a simple backdrop, even in winter.

4 This Corten water feature forms a bold centerpiece within a classic hedge, giving it a contemporary update.

5 This stone wall interrupts the hedge line, creating a backdrop for the pots in the foreground.

FURNITURE

Some people find it easier to assemble good-looking and comfortable spaces inside a house than outside, yet much of the thought process is the same. You could go cheap and cheerful or decide to invest in pieces that will stand the test of time. Before you buy, think carefully about the exact purpose of your tables, chairs, and benches and how they will fit in with your design.

What is your seating area for? A place for contemplation, a social zone, or just somewhere to sit for a quick coffee?

1 Small can be beautiful—make the most of a courtyard suntrap with a simple little table and chairs.

2 Built-in furniture maximizes seating while taking up less space. A mix of stone and wood adds interest, while the central firepit provides a focal point.

3 Furniture should feel part of your design, echoing shapes and materials used elsewhere. Seating nestled among plants offers an alternative view of the borders.

4 A single, elegant piece of furniture such as this carved bench can be a statement feature as well as a seating area.

5 A covered seating area creates an inviting, roomlike feel and offers shade or shelter from the elements.

WATER FEATURES

There's a hypnotic, magical quality to water—just like firelight—that holds people's attention. If you include water within your garden design, think about what sort of effect you want it to have. Do you want a calm, meditative mirror, perhaps, or a natural haven for wildlife, or is it to add movement and sound to your space?

1 **This sleek steel rill** combines movement, sound, and reflection as it carries water through planting down into a deep pool.

2 **A simple, plant-filled tank** is a practical option that provides water storage for use in the greenhouse as well as a haven for passing wildlife.

3 **This trio of spouts** sits at the rear of a seating area, creating a closer connection to the water and providing a lovely focal point.

4 **A free-standing reflection bowl** mirrors the sky and the plants around it, bringing a sense of calm and stillness to the space.

5 **A waterfall wall** brings this boundary to life. The widely spaced stepping stones draw the eye across the pebble-strewn pool below.

6 **A naturalistic pond** creates an oasis for water plants and wildlife. Three simple water spouts add a sense of movement and sound, creating a focal point within the feature.

LIGHTING

Whether practical or decorative, lighting adds enormous value to a garden. You may need to illuminate pathways, steps, and entrances for safety reasons, but even if your lights are purely decorative, make sure you consider their position, angle, quality, color, and intensity.

> *On days when it's dark by the time you return home, even minimal lighting just out the kitchen window will put a smile on your face.*

3

4

1 Simple tealights flickering in the night help to create a mood.

2 Colored lighting makes a change from traditional white lights and highlights planting and trees. Don't go overboard, though. Remember: less is more.

3 Lighting adds atmosphere to a garden at nighttime and means you can enjoy your space long after sunset.

4 You don't need to light your whole garden. Focus on steps and pathways to make moving around your garden easier. Uplighting can also be used to create a real focal pull.

FEATURES

A decorative feature, whether a sculpture, an urn, or other ornament, can bring a little bit of personality into your garden. Used as focal point, a feature can distract the eye or simply introduce a little fun. It doesn't need to be a big statement piece—a personal memento can have impact by bringing treasured memories into your garden.

1

2

> *Make your garden a more personal and unique space by including decorative items that are meaningful to you and your family.*

1 Personalize your garden by choosing features with artistic or sentimental value.

2 A real, functioning beehive makes a memorable and eye-catching addition to this lawn.

3 This elegant sculpture picks up the natural light in this woodland setting and looks at home surrounded by plants.

4 The form and tactile character of sculpture can draw the eye through the garden.

5 A bird table inspired by stem shapes sits comfortably in the planting.

STRUCTURES

You may want an outside office, a storage shed, or simply a pergola to add height in your garden, but if you define clearly what it's for, you can make the right design decisions about it. Take time to figure out the best position for it and the materials that will tie it to the space. You're aiming for a sense of continuity and cohesion.

1 A greenhouse can be more than just a practical structure; it can also be decorative and echo the architecture and colors of the main house.

2 A covered seating area next to still water provides a peaceful spot for contemplation and enjoying the view.

3 This steel archway feels light and welcoming. Growing fruit over the arch adds character as well as produces a wonderful harvest.

4 Rope swags create a glorious framework for roses or other climbers. Their sweeping shape gives a softer look than wood beams.

5 A covered dining area helps you make the most of your garden. A green roof not only adds an air of fun but also offers an unconventional wildlife habitat.

BUILD A MOOD BOARD

Mood boards help you figure out what you like and don't like and help you pin down the themes you want to capture in your finished design. They are a fun part of the creative process and will help you think about your design in a different way. Don't worry about what they look like—this isn't an art show.

> " *A moodboard will help you make clear decisions when you're choosing and sourcing materials so you don't get overwhelmed by all the choices.* "

WHY MAKE A MOOD BOARD?

You might think that a mood board seems unnecessary or time-consuming, but if you do create one, you will reap the benefits in the long run.

- **A sense of unity:** a mood board will help you focus in on your garden's overall theme, and identify anything that doesn't seem to fit.

- **Contrast and emphasis:** as you whittle down your list of themes, you might notice two contrasting themes that you can deliberately play off one another to create or emphasize a focal point.

- **Clear decision-making:** by referring back to your mood board when choosing and sourcing materials, you can avoid becoming overwhelmed by all the different possibilities.

- **Cost-effective:** if you have a clear range of ideas with which to work, you won't make expensive mistakes by buying random things that don't add to the overall theme.

❶ GATHER MATERIAL

Collect source material that you like, such as pictures from magazines and Pinterest, paint swatches, fabric samples, or natural objects. It might be the texture, color, pattern, or shape that catches your eye. You don't have to do it all at once—you can add to the collection over time if you prefer.

2 WORD ASSOCIATION

Note down words that describe the moods or feelings the different items evoke. Use pieces of paper or sticky notes. For example, does a texture convey serenity or tranquility? Does a color seem energizing, or is it cool and calm? Is a shape sinuous and flowing or angular and spiky? Write down the words that resonate for you.

4 ZONE IN

Pick your strongest themes and place the visuals together. Do they work together as a unified whole? If something looks odd, swap in one or two other themes in its place, until you get a few that you think work well together. Limiting yourself to a few themes will give you a stronger, cleaner design than trying to work with everything on your mood board. When you're happy with your themes, your mood board is ready. Keep it safe, referring back to it often as you continue through the design process.

3 LOOK FOR THEMES

Take a good look at what you've gathered. Are there themes in the mood words and the visuals you could use in your design? Lay them all out, play around with them, and regroup them to see if any other themes emerge.

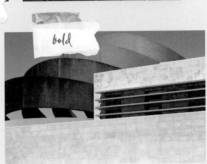

Ask *ADAM*

Looking for a style for your garden is a great opportunity to browse through lots of lovely images to decide which ones you really like and why. If you feel overwhelmed by choices, keep things simple by narrowing your search. Here are a few questions I'm often asked.

Gardens created in an ad hoc way are rarely as

beautiful or practical

as those that have had a lot of thought, love, and attention to detail put into them.

Q I'M WORRIED MY TASTE IS **TOO ECLECTIC**

To begin with, when you're first collecting images for your mood board, don't worry whether your taste looks too random and you like lots of very different styles. The aim is to look at everything, then edit back the list to those things you really love and that will work together to create an overall scheme. Often an underlying theme will reveal itself. If not, simply organize things in a hierarchy of your favorites and be prepared to play with different themes until they work as a whole.

QUICK FIX

KEEP THINGS **SIMPLE**

Using too many materials is a common mistake, even among professional esigners who can get drawn into the trap. In any garden area of any size it pays to keep things simple. When you're designing, think about your use of hard landscaping materials and how their colors work together, as well as the overall look or style you want to achieve. As a rule, I use two, three, or four materials. Once I've designed a garden, I'll go back and see if I can pare anything back a bit more.

*The point of putting together images for your **mood board** is to **discover** the kinds of things that can make your **heart skip a beat.***

WHAT KINDS OF IMAGES SHOULD I CHOOSE?

When you're putting together images for the generic mood board that will help you discover an overall aesthetic for your garden area, don't worry about getting tons of images. It's great if you choose pictures that relate directly to gardens and the outdoors, but don't be afraid to include something like a piece of music, a saying, or an image of people that sums up the mood you want to create. The point is to discover the kinds of things you are strongly drawn to and that can make your heart skip a beat.

HOW MANY GARDEN STYLES ARE THERE?

There are many different labels for garden and planting styles—cottage, terrace, tropical, and so on—but they are all open to interpretation. In reality, the best style will be the one that's unique to you. Focus on creating a garden you really love, where the various elements in it work together as a whole rather than as a random mixture of details.

WHAT **TERRACE** MATERIAL SHOULD I USE?

When choosing elements for hard landscaping you'll find that the deciding factors are cost, practicality, and aesthetics. For hard surfaces close to the house, practicality is key. Take a look at my advice on pp62–63 when weighing the different practicalities; the materials information on pp58–59 may also guide your final choice.

KEY
Knowledge

○ Collecting images of things you like is a good way of exploring different ways your garden area could look and feel.

○ Don't limit yourself to images of gardens or plants. Consider anything you are attracted to and figure out what aesthetics, colors, and moods you like best.

○ Think about words that conjure up the sort of emotional response you'd like your garden area bring about. Keep this in mind as you fine-tune your image selection.

○ When you've edited down your choices, start grouping them together into categories of the strongest themes. This will help you imagine even more clearly what your design is going to look like.

DESIGN

CHOOSE YOUR PLANTS

> *The best way to tackle planting design is to break the process into small stepping stones.*

INTRODUCTION

For me, plants are the stars of the garden area. They tell the story of the seasons, encourage wildlife, and bring your garden alive. Design-wise, plants add structure, texture, movement, sound, and color. They can also soften lines, create focal points, and help create a particular mood.

Most of us have had moments when we've felt overwhelmed by the vast number of plants out there. Good plant knowledge is something that evolves over time, and even the experts don't know them all! To be absolutely honest, you probably won't get your planting right the first time—I don't always get it right, and I've been doing this for decades.

Planting design is a big subject, and the best way to tackle it is to break the process down. In this section, I'll first help you identify your own "planting style," before showing you how to understand the layers of planting in your garden. I'll take you through some of my favorite plants and give you examples of how I develop my ideas for borders for different soil and conditions. Finally, I'll help you narrow down your plant wish list into a practical and beautiful assortment of plants with which you can go on to create your perfect borders.

> *Just like a painter uses a palette, you'll be using your palette of plants to create lovely compositions in your garden.*

FIND YOUR PLANTING STYLE

If you're new to planting design, words like "rhythm" or "repetition" might sound daunting, but planting doesn't need to be complicated. If you do a little research and prepare a mood board, you'll be able to create a garden that captures your individual personality.

Although there are lots of defined planting styles, for me, a garden should reflect someone's personality. So as you are drawn around your space, don't just think about styles; it's also good to think about mood or emotion. Your planting shouldn't be dictated just by how it looks but also by how it makes you feel.

CREATING YOUR PLANTING WISH LIST

As you start to build a wish list of ideal plants for each layer (see pp92–93), keep your planting mood board handy. Planting design isn't an exact science, but as you note down the individual plants you'd like to include in your garden, your mood board will help remind you of the bigger picture—what you want your planting to achieve as a whole—and help make sure that the plants you choose will fulfill your goals.

The best way to identify your planting style is to build a mood board (see pp82–83) specifically for your soft landscaping. Books, magazines, Pinterest, and TV shows are a great starting point for planting inspiration, but don't forget to head outside, too. Check out a local garden center to see what's available. Head to a park or a local garden open day and take a look at the borders.

Whenever you spot something you like, try to figure out what it is you like about it exactly. Are the plants tumbling into each other? Do the vibrant colors catch your eye?

Do the formal lines and symmetry stand out to you?

Wherever you go, take photos and make plenty of notes. Write down words that describe the atmosphere you'd like your garden to evoke and how you want to feel when you spend time in it. You might notice one underlying theme shining through your notes or perhaps a few themes that could influence different areas of the garden area (a quiet, contemplative space that links to a lively area for entertaining, for example). The theme(s) that you finally settle on will hopefully show a little of your personality.

> *Look at the themes and moods that emerged from your mood board to determine your planting style.*

COMMON **PLANTING STYLES**

While I don't believe you need to be familiar with well-known planting styles in order to plant a garden, you may find it useful to look up a few online as part of your mood board research. Here are a few to get you started:

- Romantic
- Formal or informal
- Minimalist or maximalist
- Contemporary
- Traditional (e.g., cottage garden or English country garden)
- A particular climate (e.g., exotic, coastal, or tropical)
- Ecological/wildlife friendly

1 Strong leaf shapes work against the straight lines of the concrete edge.

2 These lavender beds evoke memories of Provence.

3 This mixed border offers more than just flowers—there's texture, movement, and a feeling of a modern English country garden.

THINK IN LAYERS

To me, the gardens that feel right are the gardens that carry layers in their planting. These bring structure and seasonal interest, but, most importantly, they also create atmosphere. Nature is the best place to learn about planting design, and plant layers occur in all climates, from tropical rain forests to temperate oak woodlands.

" Take a walk in a woodland and you'll notice how plants naturally settle into different distinct layers of growth. "

Layers usually consist of bulbs, perennials, and shrubs, with the lower canopy and upper canopy of trees above them. In reality, layers often aren't clearly separated, so you get a blurring of lines as things intermingle. It's a really interesting way to look at gardens, and it doesn't take long to notice when something is missing in a planted scene. For me, layers make planting look and feel "right." I've shared a selection of my favorite plants for each layer across the following pages as well as examples of layering in action in my case studies.

The lower canopy is a midlayer made up of small- and medium-sized trees. This may be the tallest layer in a smaller garden. Use them to play with the light and add seasonal interest, and think about their ornamental properties, too.

The shrub layer provides structure, with the shrubs like distinct little canvases around the garden that provide an architectural backdrop for your perennial layer. Shrubs bring texture, rhythm, and seasonal interest to your planting.

The perennial layer is like the froth, or a kind of "infill" for the garden, bringing contrast, texture, movement, rhythm, and seasonal color.

The bulb layer includes true bulbs, corms, rhizomes, and tubers. Bulbs are particularly useful for bringing interest in spring, although you can also get summer- and fall-flowering bulbs.

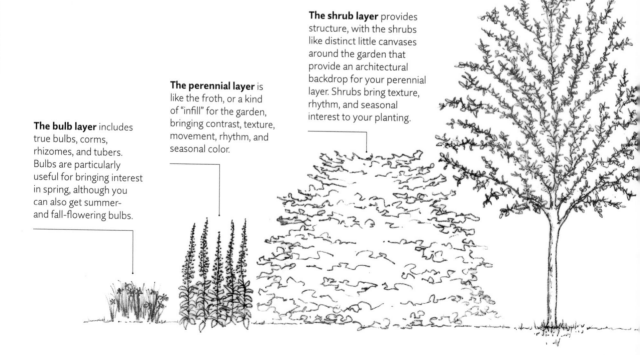

The upper canopy is the top layer of tall trees. This could be a tree you plant yourself to form a focal point, one already in the garden, or a tree in a neighbor's garden or from the surrounding landscape. You may want to "borrow" that tree in your design or plant a smaller companion tree nearby to bring it "down" into your garden (see p32). Large, shapely trees can make a garden instantly feel more mature. They can add real structure and atmosphere, help frame a view.

Planting all over the world contains different layers that make up the overall picture.

UPPER AND LOWER CANOPY OF TREES

The canopy is the biggest part of your structural planting. You may be designing around existing trees, but if you decide to plant a tree, remember that you're probably planting it for another generation. Here's a selection of trees I love to include in my own designs.

▲ *Cercidiphyllum japonicum* (Katsura tree)
One of my favourite deciduous trees, and a good multistem option, it can grow to around 39ft (12m) or more. Its small heart-shaped leaves change to shades of pink, orange, and yellow in fall. They also produce a unique burnt-sugar scent when the frosts arrive that kids love.

If you intend to plant a tree—whether it's for the upper or lower canopy—it's worth thinking practically first. How tall do you want it to grow? Do you want evergreen or deciduous? Do you want open branches to let light through or dense foliage to screen unsightly views? It's easy to be pulled in by beauty when what is really needed is an oval-shaped evergreen to block out a neighbor. Figure out what the tree's purpose is, then build up a picture of the shapes and characteristics that will work best in the space.

▲ *Acer campestre* 'Elegant' (Field maple)
This upright, compact, deciduous oval tree can grow up to 33ft (10m) tall. It's really versatile: I've grown it as pleached hedging and as a multistem. It has glorious fall color, which I enjoy seeing in our hedgerows. It's also extremely tough, tolerating most soil types, drought, and air pollution.

▲ *Betula nigra* (River birch)
This deciduous tree can grow up to 59ft (18m) tall. Its cinnamon-colored bark flakes off to reveal pinks and creams. In spring, yellow catkins are followed by soft green leaves that turn yellow in fall. It's one of the best trees for wet ground but will happily grow in drier and hotter situations.

▲ *Lagerstroemia indica* (Crepe myrtle)
A deciduous tree or shrub that can grow up to 26ft (8m), it has glossy green leaves and a multistem habit. I grow this tree in my garden, but it may be harder to grow farther north. Bright pink or red crinkly blooms appear in early fall.

◀ *Zelkova serrata* (Keaki)

A large, spreading, deciduous tree, it can reach more than 39ft (12m) in height. Not unlike *Carpinus* (hornbeam), it carries lovely smooth gray bark. It holds green oval-shaped serrated leaves that turn orange-yellow in fall, and it is tolerant of most soils.

◀ *Amelanchier lamarckii* (Serviceberry/shadbush)

This small deciduous tree is one of those plants I feel every garden should have. It can reach up to 33ft (10m), and every season brings different interest, from white star-shaped blossom in spring to berries and fiery foliage in fall. In my garden I grow this as a multistem tree, which creates a really nice shape.

◀ *Hamamelis* x *intermedia* 'Jelena' (Witch hazel)

For me, witch hazel is one of the hardest-working small trees. Its clusters of bright flowers bring a smile to my face in winter, and many have a lovely scent. This small, deciduous, slow-growing tree can reach up to 13ft (4m) tall and has great fall leaf color.

▲ *Cornus mas* (Cornelian cherry)

This small deciduous tree can grow up to 16ft (5m) tall with year-round interest and superb bark as the plant matures. In spring, it has tiny yellow flowers. In fall, the foliage turns red and deep purple and carries small, cherry-like berries.

▲ *Ginkgo biloba* (Maidenhair tree)

I do love the history of this tree—it's one of the oldest trees on Earth. It can reach up to 100ft (30m) and has unusually fan-shaped deciduous green leaves that turn brilliant yellow in fall.

▲ *Malus* 'Evereste' (Crab apple)

A small deciduous tree that can grow up to 23ft (7m), with pretty spring blossom that starts as pink buds and opens to white. It bears lots of lovely small ornamental yellow-orange fruit in fall. It is one of the most disease-resistant crabs.

MIDLAYER SHRUBS

Shrubs seemed to fall out of fashion for a time. However, I've always thought that gardens, particularly average-sized gardens, really benefit from the structure and interest shrubs provide throughout the year. Here are a few varieties I like to use, but it's good to explore other options, too!

In addition to structure, shrubs make a good backdrop for perennials and bulbs. They can also bring a lovely sense of rhythm to a space. The biggest mistake people make when buying shrubs is to buy one of this and a couple of that and plant them too close together. A few years later, the shrubs have become one big messy mass. Think about what you want the shrub to bring to your space—the shape, size, foliage, flowers, scent—and how they might work with your other plants. Then make a list of those best suited to the conditions in which you want to plant them.

▲ *Aesculus parviflora* **(Dwarf buckeye)**
This is an underused shrub that deserves more attention. It has large, dramatic deciduous leaves, which turn shades of yellow in fall, and unusual white flowers in summer. It grows up to only about 10ft (3m) high but has a suckering habit so can get quite wide. I've seen it as an elegant multistem with a clipped canopy.

▲ *Salix elaeagnos* **subsp.** *angustifolia* **(Narrow-leaved olive willow)**
I've often used this willow in show gardens because of its great texture and shape. The deciduous leaves are very fine, like rosemary, so although this shrub can grow up to 10ft (3m), it never looks too bulky.

▲ *Aronia x prunifolia* **(Chokecherry)**
This deciduous shrub, which can grow up to 13ft (4m) tall, is another plant that just fascinates me all year. Clusters of small white flowers with pale pink stamens are followed by jet-black fruit, which are edible. Its fall color is fantastic.

▲ *Hydrangea arborescens* **'Pink Annabelle'**
I came across this deciduous shrub a few years ago. A really delicate pink, it grows up to 8ft (2.5m) high. It produces masses of large long-lasting flower heads.

◀ *Viburnum plicatum* f. *tomentosum* **'Mariesii' (Japanese snowball)**
This spring-flowering deciduous shrub, with large lacecaps of white flowers, is perfect. The flowers are followed by berries, which add another layer of interest. I do like this shrub. Even when it loses its leaves, it still has a really strong form. It can reach up to 10ft (3m) in height and has distinctive horizontal, tiered branches. It looks great planted between a more formal and informal space.

◀ *Sarcococca hookeriana* var. *digyna* **(Sweet box)**
Around 5ft (1.5m) tall, this evergreen shrub (or small hedge) has dark green, lance-shaped leaves and tiny, creamy-white highly scented flowers, followed by glossy black berries.

> " *The right choice of shrubs can add structure and a great sense of rhythm to a garden.* "

▲ *Cornus alba* **'Kesselringii' (Tatarian dogwood)**
The stems of this hardy 10ft (3m) deciduous shrub turn to blackish-purple in winter, and its green leaves turn reddish in fall. It can be cut back hard to encourage new growth.

▲ *Rosa macrophylla* **'Doncasterii' (Rose)**
This deciduous rose grows up to 5ft (1.5m) tall and has a long, arching stem of growth, with bright pink flowers followed by long red hips. I fell in love with it after using it in a Chelsea garden.

▲ *Phlomis russeliana* **(Turkish sage)**
This vigorous, spreading, deciduous plant offers lots of interest all year, from its heart-shaped foliage to its 3ft (1m) tall stems with whorls of pale yellow flowers. The seed heads look particularly beautiful covered in frost.

▲ *Mahonia* x *media* **'Winter Sun' (Oregon grape)**
A plant I've fallen back in love with, it has rosettes of glossy, dark green, evergreen leaves and spikes of yellow highly fragrant flowers. It grows up to 16ft (5m) in height.

▲ *Philadelphus* **'Belle Etoile' (Mock orange)**
Who doesn't love the scent of a mock orange? In late spring and summer, this compact shrub, which grows up to 4ft (1.2m) tall, is covered with an abundance of highly scented white flowers. The deep green oval leaves are deciduous.

PERENNIALS

These plants are, for me, what really brings your garden alive through the growing season. What's lovely about perennials, including some of my favorites here, is that it's easy to chop and change them until you've created combinations you're happy with. With so many to choose from, the fun feels endless.

Although trees and shrubs are bigger, it's the perennials (the nonwoody plants) that will make up most of your planting. Perennials come in many shapes and sizes and can be hardy, half-hardy, evergreen, or deciduous—or, in the case of herbaceous perennials, die back and reappear each year. Think about when a plant looks at its best and make sure it suits your soil and climate. Inevitably, you'll make a few mistakes, but just keep playing—I still am!

▲ *Achillea 'Inca Gold'* (Yarrow)
This yarrow can grow up to 24in (60cm) tall. Its silvery-green foliage works well with other plants. Its bright orange umbels age to bronze.

Eryngium agavifolium (Sea holly) ▶
A handsome architectural plant that can grow up to 3ft (1m) tall, sea holly demands your attention with its rosettes of spiky foliage and clusters of cone-shaped gray-green flowers on stout stems.

▲ *Eutrochium purpureum* (Joe Pye weed)
This is a good structural plant that works in most soils in either sun or part shade. Slightly domed flowers add a delicate touch to this giant, which can reach up to 7ft (2m) tall.

▲ *Foeniculum vulgare* (Fennel)
Fennel is such a valuable plant. With its fine, soft foliage, it adds up to 6ft (1.8m) of height and interesting texture to a border. You do need room for this one to work its magic.

▲ *Limonium platyphyllum* 'Violetta' (Sea lavender)
This plant has a coastal feel to it and works well in a gravel garden. Growing up to 20in (50cm) tall, it has purple flowers above a clump of evergreen leaves that can change color in the sun.

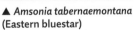

**◀ Panicum virgatum 'Heavy Metal'
(Switch grass)**
An impressive ornamental grass with upright metallic blue leaves up to 5ft (1.5m) tall that turn yellow in fall. A haze of pink-tinted flowers appear in summer. For me, this one really earns its place in the fall months.

▲ Valeriana pyrenaica (Capon's tail grass)
This lovely plant makes a good statement in a border without being too loud or overpowering. The heart-shaped leaves are a nice feature earlier in the year, but the clouds of tiny flowers on stems that grow up to 3ft (1m) tall are the real stars.

◀ Dianthus carthusianorum (German pink)
This tall-stemmed pink has single magenta flowers above grassy foliage. Reaching up to 16in (40cm) tall, it looks great in borders, in long grass, or in a meadowlike planting.

> " *What's lovely about perennials is that you can chop and change them about until you've created combinations you're happy with.* "

**▲ Amsonia tabernaemontana
(Eastern bluestar)**
An underused plant that grows up to 31in (80cm) tall, it has star-shaped ice-blue flowers and glossy green leaves that turn yellow in fall.

▲ Centaurea 'Jordy' (Knapweed)
A hard worker, knapweed will keep on flowering, and its plum-colored thistlelike flowers work well in borders with an array of colors. It can grow up to 20in (50cm) tall.

▲ Melica uniflora (Wood melick)
This elegant, understated plant has delicate sprays of pale golden-brown seed heads above mounds of soft green grass that can reach 24in (60cm) tall. It can survive even in dry shade.

▲ *Aruncus* 'Horatio' (Goat's beard)
This clump-forming plant has wonderful fine foliage that is easy to plant with. Its 3ft (1m) tall stems bear small plumes of creamy-white flowers that fade to bronze and provide a good fall splash, too.

Iris sibirica 'Tamberg' (Siberian iris) ▶
Perhaps my favorite group of iris to work with, Siberian iris are very reliable. They grow up to 35in (90cm) tall and have pale blue butterfly-like flowers.

▲ *Molinia caerulea* subsp. *caerulea* 'Edith Dudszus' (Purple moor grass)
This tuft-forming grass carries its flowers high above its foliage, which grows up to 3ft (1m). The leaves are a beautiful color in fall, and the flowers hold well until late in the year.

▲ *Persicaria affinis* 'Superba'
This semievergreen plant forms a wide mat of slim, midgreen, up to 12in (30cm) tall leaves and short spikes of pale pink flowers that darken as they age. I love to use it as ground cover and work other plants around it.

▲ *Rodgersia pinnata* 'Elegans'
This handsome, architectural plant grows up to 4ft (1.2m) tall. The leaves carry a rough texture and color well in fall. It produces handfuls of small, star-shaped pink to creamy-white flowers in summer.

Astrantia major 'Large White' (Masterwort)

At around 24in (60cm) tall, astrantias are a great midheight plant for sun or part shade. The white flowers with green tips seem to last forever.

Aquilegia chrysantha 'Yellow Queen' (Columbine)

This graceful plant has unusually long spurs on large lemon-yellow flowers, and grows up to 36in (90cm) tall.

Dryopteris filix-mas (Male fern)

This fern grows in damp woodlands. It has a great shape and texture and looks great planted next to hostas and grasses. If I have space, I like to use it en masse and play around with contrasting foliage. It can grow up to 3ft (1m) tall.

Euphorbia amygdaloides var. *robbiae* (Wood spurge)

This handsome, hardworking perennial can be a little invasive, so I tend to use it in dry shade only. It bears dark, glossy evergreen leaves, grows up to 24in (60cm) tall, and produces masses of showy, long-lasting, lime-green flowers in late spring.

Geranium nodosum (Knotted cranesbill)

You can find geraniums for most conditions. This pretty plant grows to around 16in (40cm) tall and produces an endless display of pinky-lilac flowers.

Iris foetidissima (Stinking iris)

One of only two iris native to the UK, it has glossy green leaves and small purple-yellow flowers, forms large seed pods, and can grow up to 31in (80cm) tall. It's great for shade but is toxic if eaten by people or pets.

Matteuccia struthiopteris (Ostrich fern)

I'm not sure there is anything more beautiful in spring than this deciduous, delicate fern unfurling itself into a shuttlecock shape. Reaching up to 5ft (1.5m), it adds tall structure.

BULBS

Bulbs are one of the easiest plants to grow. They look great planted in the beds and borders and naturalized in wilder areas. I also love having big containers of them close to the house in early spring so I can enjoy them if the weather is still too cold to venture out much into the garden. Here are some I really like.

▲ *Allium sphaerocephalon* (Drumstick allium)
This sun-loving plant can grow up to 2½ft (80cm) tall. It has small purple and green flower heads on strong wiry stems that add lovely color and movement to a border. I like to plant them with grasses and drift them through my borders as if they had self-seeded there naturally.

The word "bulb" is often used for any plant with an underground food storage organ, but botanically, bulbs should be categorized as true bulbs, tubers, corms, and rhizomes. Whatever you want to call them, they bring lovely seasonal surprises, and many are scented, too. Bulbs come in all sorts of heights and shapes and are useful for bringing a splash of color into the garden early in the year before some of the herbaceous plants have got going. Some types of bulbs will come up year after year and even seed themselves around, while others will need replacing every season.

▲ *Narcissus 'Minnow'* (Daffodil)
This miniature daffodil can reach 8in (20cm) tall, and has clusters of 2–4 pale yellow flowers per stem. It prefers a sunny, sheltered spot and often spreads well, with bigger displays each year.

▲ *Cyclamen coum* (Cyclamen)
These tough yet dainty pink flowers can grow up to 4in (10cm) tall and look beautiful above the rounded green leaves. They make perfect companions for snowdrops and hellebores.

▲ *Camassia quamash* (Common camassia)
With spikes that can grow up to 32in (80cm) covered in blue star-shaped flowers, this camassia is perfect for planting in long grass. They look just as happy in a meadow or border.

◀ *Eranthis hyemalis* (Winter aconite)
Able to grow up to only 4in (10cm) tall, what this tiny little plant lacks in size it makes up for in beauty. Best grown in part shade, it is an effective way of adding pops of color to a border. It's a cracker and never lets me down.

▼ *Crocus tommasinianus* (Crocus)
I tend to use multiple varieties of this crocus bulb to achieve a multicolored "dolly-mixture" look, which creates a real focal pull in early spring. I also like to naturalize in lawns as they are shorter, growing up to 4in (10cm), and don't look scruffy after flowering like some bulbs.

▼ *Iris reticulata* (Iris)
As these iris are only small, reaching up to 4in (10cm), I like to plant groups of them in terra-cotta pots. When the shoots poke through the soil in late winter, I move the pots somewhere that I pass daily so I can enjoy their cheery flowers. They also work well in gravel plantings.

▲ *Tulipa sylvestris* (Wild/species tulip)
This native tulip is one of my favorite bulbs. It has a scent that pulls you to your knees. At 12in (30cm) tall, I plant it in the grass in my orchard as it's less showy than other tulips.

▲ *Galanthus nivalis* (Snowdrop)
I love snowdrops. It's always a nice surprise when the timeless white flowers appear. I really like them in a woodland setting, and at only 6in (15cm) tall, they're great planted in grass and left to naturalize.

▲ *Narcissus pseudonarcissus* (Wild daffodil)
How can you not fall in love with the dainty native British wildflower that inspired Wordsworth? Reaching 10in (25cm) tall, it's an easy daffodil to grow and will spread by seed.

CLIMBERS AND WALL SHRUBS

Climbing plants are really useful for covering boundary walls and fences and for bringing interest to archways and pergolas. Some are evergreen, while others are deciduous—I have favorites of both types, as you can see below.

Using climbers and wall shrubs to cover vertical surfaces is a great way to soften the hard landscaping elements of your garden and give it more atmosphere. Climbers tend to have a small footprint and will either be self-clinging (attaching and supporting themselves) or need a structure such as trellis or a network of wires that they can be attached to. Climbing shrubs, also called wall shrubs, usually need more space to grow and don't climb naturally so need to be encouraged to grow upward with careful pruning and tying in to some sort of support.

▲ *Rosa 'Paul's Himalayan Musk' (Rose)*
One of the prettiest deciduous rambling roses, this is particularly vigorous with clusters of dainty, double, pale pink highly scented flowers. Before you plant, make sure you have put in a support for it to climb up.

▲ *Trachelospermum jasminoides (Star jasmine)*
The neat glossy evergreen foliage of this elegant climber is really useful for covering walls. In summer, it is covered in clusters of highly fragrant, star-shaped white flowers. It has a twining habit and works well on a wired wall. Apart from a little early help, it doesn't need much training or pruning.

▲ *Vitis coignetiae (Crimson glory vine)*
The dark green leaves of this deciduous climber turn into spectacular shades of red, orange, and crimson-purple in fall. It is fast growing and dense so is best planted by itself as it can smother other plants. You will need to tie in branches to some sort of trellis or support.

▲ *Wisteria sinensis (Asian wisteria)*
Every year, I look forward to my wisteria flowering—seeing the long bunches of buds fill out then burst open to create a wash of pale purple. This quick-growing deciduous plant is great trained against a wall or over a strong pergola.

◄ *Lonicera x tellmanniana* (Tellman's honeysuckle)
This unusual honeysuckle has clusters of unscented burnt-amber flowers flushed with red. A deciduous climber, it has a dense, twining habit that needs to be supported.

◄ *Chaenomeles speciosa* 'Geisha Girl' (Japanese quince)
Flowering later than other ornamental quince, 'Geisha Girl' is deciduous, has pretty apricot-pink flowers, and a compact habit. The fruits can hold into winter.

◄ *Hydrangea anomala* subsp. *petiolaris* (Climbing hydrangea)
A long-time favorite, this self-climbing deciduous plant has a sea of green leaves and beautiful white flowers in the summer. Even the winter skeleton is great. It's happy in a shady spot.

▲ *Garrya elliptica* (Silk tassel bush)
A tall evergreen shrub with glossy dark green leaves and a profusion of long, silky, gray-green catkins, which grow up to 12in (30cm) long and are particularly impressive on male plants. It prefers a sheltered spot out of cold winds.

▲ *Rosa filipes* 'Kiftsgate'
This deciduous rambling rose can grow to more than 33ft (10m) in height. The fragrant white flowers appear in late summer, followed by tiny orange-red hips. If you have an old shed to cover or a large tree it can climb up, it works wonders.

▲ *Parthenocissus henryana* (Chinese Virginia creeper)
With velvety, silver-veined, deeply divided leaves that turn crimson in fall, this deciduous climber can grow to more than 33ft (10m) in height. The fall color is most dramatic if you grow it in a sunny spot.

ANNUALS, BIENNIALS, AND SHORT-LIVED PLANTS

Plants that live for just a year or two are a great way of experimenting with shape and color without the commitment of more permanent planting. Of course the choice is huge, but here is a small selection of my favorites.

Many annuals, biennials, and other limited-life-span varieties can be grown from seed, making them a cheap and cheerful method of planting. I often use them to fill spaces in my planting—for instance, if I'm developing a border over a long period of time or if I'm waiting for a shrub to fill out. I also like to overseed areas with biennials then let them do their own thing, as it makes a plant scheme look more relaxed.

▲ *Nicotiana sylvestris* (Flowering tobacco)
With tall stems, lush leaves, and clusters of dramatic tubular white flowers, this plant is a real eye-catcher. Its heavy night scent will really stop you in your tracks, too.

Verbascum phoeniceum hybrids (Mullein) ▶
With tall, long-flowering spikes that add great vertical accents in a border, this seed mix produces a lovely range of colors.

▲ *Ammi majus*
(Queen Anne's lace)
A sort of refined cow parsley, Queen Anne's lace forms a prolific froth of delicate white umbels above fine fernlike foliage.

▲ *Digitalis purpurea* (Foxglove)
A common sight in the wild, foxgloves equally make great garden plants and are beautiful drifted through a woodland. They are biennial, so plant two years in a row to get a succession of flowers.

▲ *Hordeum jubatum*
(Foxtail barley)
With long, silvery seed heads that fade to pink, this short-lived grass deserves to be planted more! I love using it in gravel plantings. Sow in either spring or fall.

▲ *Nigella damascena* 'Miss Jekyll'
(Love-in-a-mist)
This pretty sky-blue annual with wispy foliage is great for a cottage garden. Like sweet peas (right), it makes a really fantastic cut flower.

◄ *Verbena bonariensis* (Tall verbena)

Tall verbena's wiry stems, topped with small purple flowers, can reach up to 6ft (1.8m). Because of its delicate structure, this sometimes short-lived perennial can be fitted in pretty much anywhere in a border and will self-seed freely if happy.

▼ *Angelica archangelica* (Angelica)

This plant has a really strong architectural form, which is why I like to grow it in my garden. Don't be afraid to place one of these at the front of a border. After the flowers have faded, the seed heads still provide great structure.

▼ *Lunaria annua* (Honesty)

Honesty has pretty lilac or white flowers in late spring but is probably best known for its translucent, silvery-golden seed cases. It's a biennial so worth sowing every couple of years to start with, after which it will self-seed.

▲ *Lathyrus odoratus* (Sweet pea)

Sweet peas are easy to grow from seed, although you can cheat and buy seedlings in spring. They come in many colors and most are highly scented.

▲ *Papaver commutatum* 'Ladybird' (Ladybird poppy)

A large brilliant red poppy with a black splodge at the base of the petals, this is a stunner and just makes me smile. It's easy to grow, too—just scatter seeds in spring.

▲ *Orlaya grandiflora* (White lace flower)

A hardy annual with delicate lacy white flowers, it can be sown in spring but produces bigger flowers if sown in fall and overwintered.

▲ *Briza maxima* (Quaking grass)

A great low-growing grass that I like to plant near the front of a border. It has fine long leaves and locket-shaped flowers, which react beautifully with the light and add movement in the slightest breeze.

HEDGES

If you have enough space to plant them, hedges are the best way of making your boundaries disappear. As well as hiding existing fences, they can create boundaries in their own right. Depending on what you choose, a hedge can form a great backdrop to offset planting in a border and can also make a fantastic habitat for birds.

▲ *Rosa canina* (Dog rose)
The dog rose is deciduous, fast growing, and tolerant of lots of different conditions. It forms an impenetrable thorny thicket and works well as part of a mixed hedge. The pretty pink flowers are followed by striking red oval-shaped hips.

If you decide to plant a hedge, there a quite a few options to choose from—including the ones I've mentioned here—so think carefully about what you want out of it. Will it hide a fence, be a foil for your planting schemes, or attract wildlife? Think also about the aesthetics. Do you want the constant backdrop of an evergreen or something deciduous that changes with the season? Whatever you choose, make sure it suits your growing conditions. It's always worth looking what hedging plants grow well in your local area. Finally, think about height and make sure it isn't going to cast too much of a shadow on your or your neighbor's garden.

▲ *Buxus sempervirens* (Boxwood)
With its small leaf and dense habit, boxwood makes a great evergreen hedging plant. To help avoid blight problems, make sure it has good air circulation and is well fed.

▲ *Acer campestre* (Field maple)
Field maple is a great option for a deciduous stand-alone or mixed native hedge. The leaves turn bright buttery yellow in fall, and it grows well in most conditions.

▲ *Taxus baccata* (Yew)
A classic evergreen hedge, yew looks comfortable in a traditional or contemporary garden. It's not as slow growing as you might think. The fruit is poisonous to humans.

◀ *Carpinus betulus* (Hornbeam)
Although it looks very similar to beech, hornbeam is a hardworking plant and far better at tolerating poor soil and exposed conditions. Although it loses slightly more leaves than beech in winter, they do turn lovely shades of yellow to orange in fall. It is a very flexible plant.

▼ *Fagus sylvatica* (Beech)
Common beech stands out for its beautiful soft-green leaves with wavy edges. Strictly speaking, it is classed as deciduous, but it holds on to its old leaves until the new ones push them off. It turns a fantastic copper color in fall.

▼ *Prunus spinosa* (Blackthorn)
Blackthorn forms a dense, prickly deciduous hedge with vicious thorns and masses of small white flowers early in the year, as well as sloes in fall. It works well as part of a mixed hedge too.

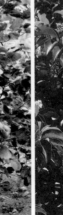

▲ *Crataegus monogyna* (Hawthorn)
Hawthorn will grow almost anywhere. It has small, glossy deciduous leaves, scented white flowers, and lovely red berries. It's a great plant for nesting birds.

▲ *Corylus avellana* (Hazel)
Hazel is a lovely bushy deciduous hedge that creates an informal but dense screen. It grows well in most situations and is tolerant of exposed sites. You might also get a good crop of nuts!

▲ *Prunus lusitanica* (Portuguese laurel)
With a smaller, darker leaf than the usual laurel, this evergreen makes a smart, dense, fast-growing hedge. It works well as a backdrop for planting.

A MIXED BORDER

The ideal situation for many plants is one that is not too sunny or shady and not too dry or damp. With average conditions, you can have fun with a huge range of plants. Here, I've chosen some of my favorites that I think work together beautifully throughout the layers, and overleaf I talk you through how I put them together.

This design offers a rich palette of interest that will last through the summer and into fall.

MY DESIGN

This particular combination of plants provides color, texture, good interest, and rhythm that plays throughout the border. For the bulb layer, not shown on the plan, I've chosen an early *Narcissus* and a dramatic drumstick *Allium*.

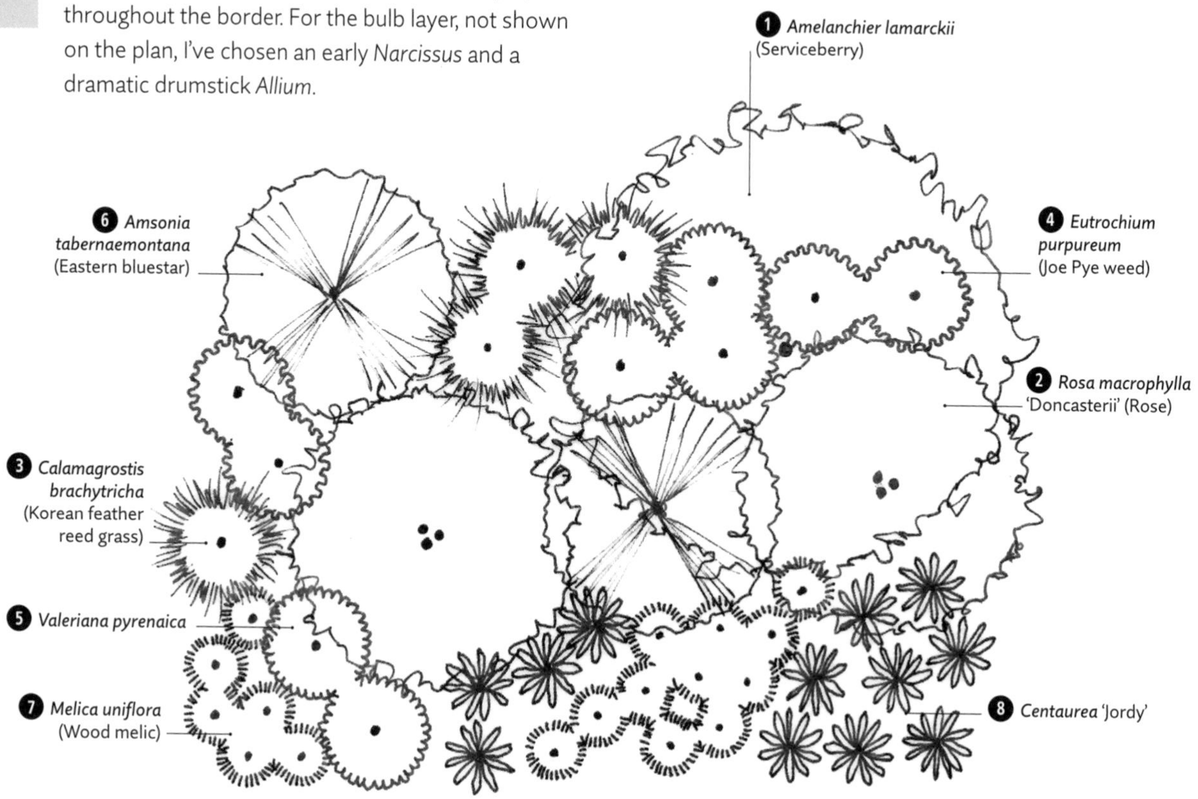

1 *Amelanchier lamarckii* (Serviceberry)

6 *Amsonia tabernaemontana* (Eastern bluestar)

4 *Eutrochium purpureum* (Joe Pye weed)

2 *Rosa macrophylla* 'Doncasterii' (Rose)

3 *Calamagrostis brachytricha* (Korean feather reed grass)

5 *Valeriana pyrenaica*

7 *Melica uniflora* (Wood melic)

8 *Centaurea* 'Jordy'

OVERHEAD PLANTING PLAN

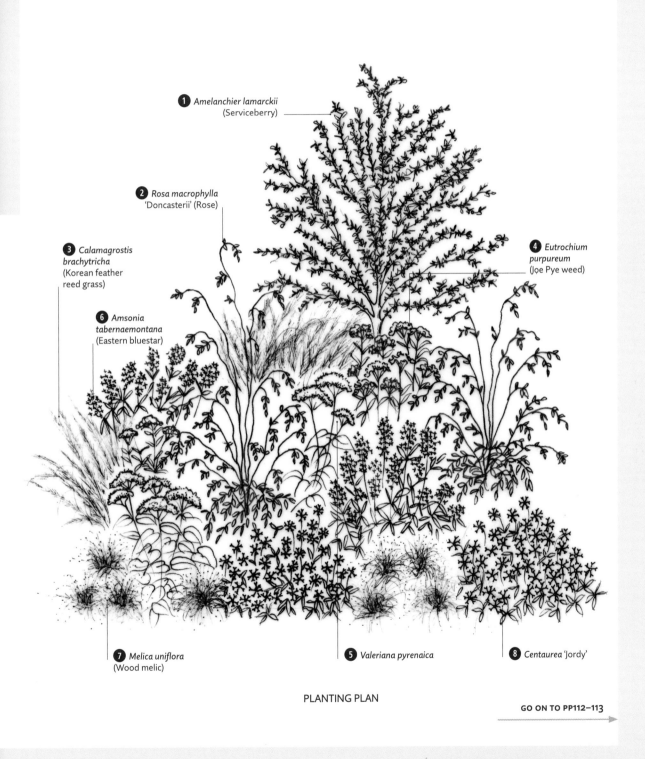

1 *Amelanchier lamarckii* (Serviceberry)

2 *Rosa macrophylla* 'Doncasterii' (Rose)

3 *Calamagrostis brachytricha* (Korean feather reed grass)

4 *Eutrochium purpureum* (Joe Pye weed)

6 *Amsonia tabernaemontana* (Eastern bluestar)

7 *Melica uniflora* (Wood melic)

5 *Valeriana pyrenaica*

8 *Centaurea* 'Jordy'

PLANTING PLAN

GO ON TO PP112–113

HOW I WORK

Here, I start with the lower canopy, using it to provide structure and create a springboard for the rest of the border. I'm looking for a rhythm of forms across and through the bed and contrasting foliage and form to create interest.

Amelanchier lamarckii (Serviceberry)

❶ FIRST, THE CANOPY

I use a small *Amelanchier* (about 9–16ft/3–5m), which is probably my favorite small tree, not just for its shape—which is open in habit—but also for its leaves, which start the year with a bronze tinge and then finish with beautiful fall colors.

Next, I'm thinking about shape and form in front of the tree—I want to partly frame it but without blocking it, so light and airy planting would work. Behind the tree, I want the planting to be upright but to sit lower than the tree and provide color, form, and movement. First, then, I'm probably just looking to add three varieties of plants to start forming the next layer (see Steps 2, 3, and 4).

❷ THE NEXT LAYER

My first shrub is *Rosa* 'Doncasterii', which will grow to around 6ft (1.8m) and has a wonderful arching habit. I can picture it looking like it's growing out of the other plants, adding details and providing good form. It has lovely foliage and flowers, but my favorite feature is the flagon-shaped, orangey-red hips that carry through the winter months.

Rosa macrophylla 'Doncasterii' (Rose)

❻ ADD ANOTHER SHAPE

Most plants so far are arching or upright, so now I want to add a more rounded form to sit centrally in the scene and, at the same time, offer a contrast in leaf. The *Amsonia* does just that, with its airy and open yet tidy willowlike foliage. After the exquisite pale blue starlike flowers, the plant finishes the year with a golden flush as the leaves take on their fall color.

The border is now really starting to take shape, with a good blend of seasonal interest, color, texture, and form.

Amsonia tabernaemontana (Eastern bluestar)

❼ BEGIN TO ADD DETAIL

Now my attention turns to the front detail, where I want to echo a shape from the back of the space. I pinch the vertical detail from the back of the bed in the form of *Melica*, a grass that will provide rhythm, movement, and life. With its neat clumps of soft green foliage and grainlike flowers that move gracefully in a breeze, it's one of those tactile plants that really draws you in.

Melica uniflora (Wood melic)

Calamagrostis brachytricha (Korean feather reed grass)

3 ADD CONTRAST

Behind the rose, I need good interest and contrast, so I'm drawn to the strong vertical of grasses. *Calamagrostis* works well: first, because it's a robust plant with a reliable clump-forming habit with strong verticals; second, because it has fluffy purple-tinged flowers that look great in the winter months. It also works well as a filter to the rose.

4 BUILD RHYTHM

Against the grass, flat-flowered plants work well, and I also want something very upright. I'm still thinking about the back of the border, but I want the rhythm to start coming forward, so I add *Eutrochium* at the back and on the side nearer the front. It's a reliable, clump-forming plant with large, flat-domed flowers that, for me, will add a little drama to any border.

Eutrochium purpureum (Joe Pye weed)

5 MORE CONTRAST

Now I want to carry on bringing height to the front and adding some contrast in leaf form. *Valeriana* is perfect for this. Its large heart-shaped, light green leaves alone create great impact. Sitting about 3ft (1m) above the foliage are large clusters of soft pink flowers, but the seed heads, which are like fluffy clouds, are probably my favorite part of this plant.

Valeriana pyrenaica

Centaurea 'Jordy'

8 BUILD THE DETAILS

I also want to add good seasonal flowers, and I need a hardworking plant that contrasts with the others. The *Centaurea* will flower throughout summer above green-gray furry foliage. The spiderlike flowers are a deep burgundy, like a good bottle of red wine. When the flowers are finished, the seed heads provide interest for a good few months.

THE SPRING BULBS

In the bulb layer, I want to have good color early in the year as well as a plant to add detail and a sense of rhythm through the border. *Narcissus*, which blooms from March to April, provides the color. It's an old but classic variety and not like the usual daffodils. It has green-white buds from which rather elegant snowy-white, multiflower heads emerge, carrying a sweet scent.

Allium sphaerocephalon (Drumstick allium)

THE SUMMER BULBS

Lastly, I've put in a great drumstick *Allium* that I love to work through grasses. Here, I'm planting major clumps, then letting the bulbs drift from there. The egg-shaped claret flowers appear in July and August and sit above slender stems. Even when the flowers fade and dry, the plant's seed heads can still provide interest well into fall.

Narcissus 'Thalia' (Daffodil)

A SHADY BORDER

When you have shady conditions, the choice of plants that will thrive becomes a little more limited. That said, there are still lots of beautiful examples you can plant to create a lush little oasis. I've designed a planting scheme that combines texture, green tones, and light-colored flowers to really enhance a shady area.

> *I've used different foliage textures and colors to bring a dark corner to life.*

MY DESIGN

All the plants I select have to work hard, and that is particularly so for shady areas. See overleaf for how I put this combination together. The choice of bulbs was simple—winter aconites and snowdrops, planted in a large drift with a few more loosely spaced on the outer edges.

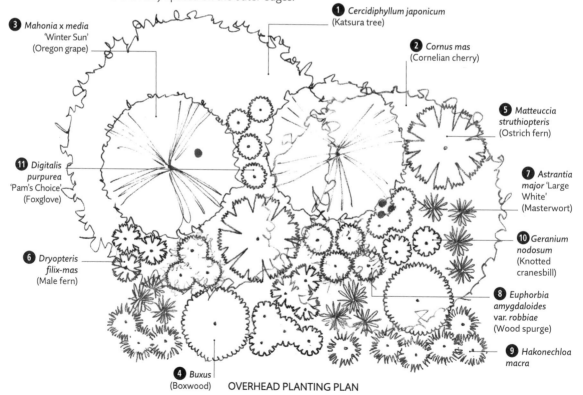

3 *Mahonia x media 'Winter Sun'* (Oregon grape)

1 *Cercidiphyllum japonicum* (Katsura tree)

2 *Cornus mas* (Cornelian cherry)

5 *Matteuccia struthiopteris* (Ostrich fern)

11 *Digitalis purpurea 'Pam's Choice'* (Foxglove)

7 *Astrantia major 'Large White'* (Masterwort)

10 *Geranium nodosum* (Knotted cranesbill)

6 *Dryopteris filix-mas* (Male fern)

8 *Euphorbia amygdaloides var. robbiae* (Wood spurge)

9 *Hakonechloa macra*

4 *Buxus* (Boxwood)

OVERHEAD PLANTING PLAN

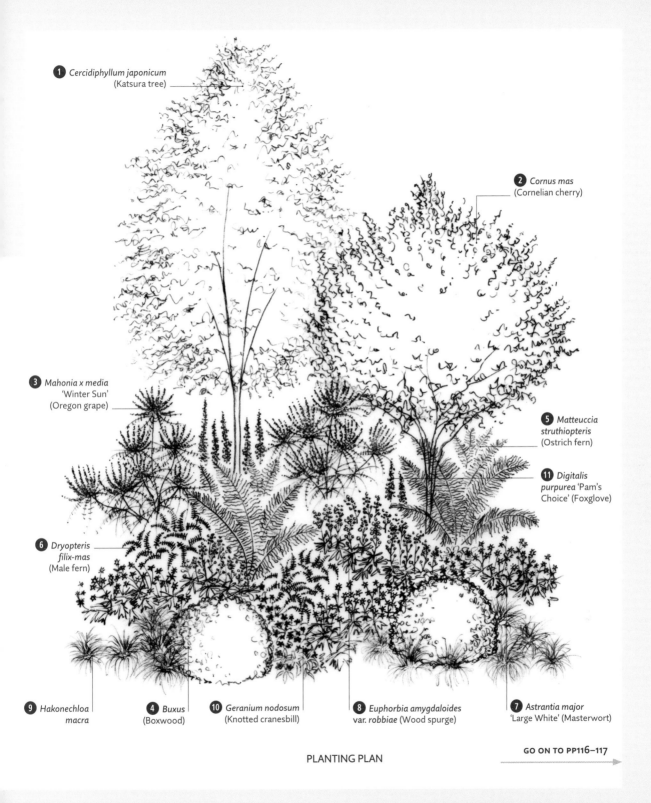

1 *Cercidiphyllum japonicum*
(Katsura tree)

2 *Cornus mas*
(Cornelian cherry)

3 *Mahonia x media*
'Winter Sun'
(Oregon grape)

5 *Matteuccia*
struthiopteris
(Ostrich fern)

11 *Digitalis*
purpurea 'Pam's
Choice' (Foxglove)

6 *Dryopteris*
filix-mas
(Male fern)

9 Hakonechloa
macra

4 *Buxus*
(Boxwood)

10 *Geranium nodosum*
(Knotted cranesbill)

8 *Euphorbia amygdaloides*
var. *robbiae* (Wood spurge)

7 *Astrantia major*
'Large White' (Masterwort)

PLANTING PLAN

GO ON TO PP116–117

HOW I WORK

As always, I work down through the layers, picking out different textures and tones and weaving in pops of color and seasonal interest.

Cercidiphyllum japonicum (Katsura tree)

1 FIRST, THE CANOPY

The multi-stem form of the *Cercidiphyllum* will restrict the maximum height of the canopy a little while adding a top layer that will provide dappled shade. This tree has very good and interesting foliage—it's crisp and fresh in spring, with a warmth to the foliage throughout summer. It also has a leaf with a lovely shape, which is easy to contrast with the next layer of planting. Fall brings fantastic color and a wonderful surprise as the frosts arrive, and the tree fills the air with the scent of burnt sugar.

Cornus mas (Cornelian cherry)

2 NEXT, THE CORNUS

The *Cornus* is a hardworking small tree that will sit well with the *Cercidiphyllum* thanks to the contrast of its green elliptical leaves. I use a multistem to get more bark, which, as it ages, has scaly plates of orange brown that give great winter interest. Small clusters of yellow flowers appear very early in the year, which are followed by dark red fruit the following fall.

6 CREATE A RHYTHM

I then add *Dryopteris* away from the *Matteuccia*, so that, although they have a different foliage color, they start to create a rhythm through the border. I also included *Dryopteris* because it is a robust old thing that never seems to let me down!

Dryopteris filix-mas (Male fern)

Astrantia major 'Large White' (Masterwort)

7 ADD COLOR

My next port of call is color—but again, the plant needs to work hard. I am adding *Astrantia* not only because of its pin-cushion flowers—which are mainly white and fade to green at the ends of the petals—but also for its lovely palmate-lobed basal leaves, which provide good contrast with the ferns.

8 GROUND COVER

In shady areas, I also think about ground cover. The *Euphorbia*, for me, comes into its own in shady spaces. The first thing I use this plant for is the foliage shape; it has long stems covered in rosettes of dark green leaves, which are usually evergreen. Above these, round sprays of acid-green flowers arrive in spring, which work really well in a dark area.

Euphorbia amygdaloides var. *robbiae* (Wood spurge)

3 ADD STRONG FORM

The *Mahonia* works well in the shrub layer as it holds a strong architectural form, which provides not only good interest with its dark spiky foliage but also a good form to plant the next layer against. As a bonus it also has good winter flowers and scent, but I'm using it here mostly for its architectural interest.

Mahonia x media 'Winter Sun' (Oregon grape)

Buxus (Boxwood)

4 VISUAL LINKS

In the shrub layer, I've used clipped *Buxus* close to the front of the border. With its evergreen foliage, it provides a visual link with the *Mahonia*, as well as strong form and a focal pull to start to build the other planting around. In a bigger border, you could use these to create rhythm and movement.

5 PROVIDE TEXTURE

My perennials form the next layer, and it's not all about the color. This is especially true in shady areas, where texture works really well, so the ferns are the first plant I add. *Matteuccia* gives height and a light lime-green foliage, which really does pop in a shady space. For me, it is a plant that is really exciting early in the year when the fronds unfurl.

Matteuccia struthiopteris (Ostrich fern)

9 VERTICAL FORM

Now I'm looking to add a little vertical form and movement. Grasses are great for this, and here I have gone for *Hakonechloa*, which is a lovely mound-forming bright green grass. It has a light airy flower in midsummer. You also get a decent fall color, and it contrasts really well with the small-leaved *Buxus*.

Hakonechloa macra

10 ADD DETAIL

Next, I'm starting to think about the front details and perhaps a little more color. The *Geranium nodosum* is a small-flowered plant that can be semievergreen. It has a sprawling habit with light green three-lobed foliage, which brings something new. It is also very long flowering—it starts in late spring, and I have seen it still going in early fall.

Geranium nodosum (Knotted cranesbill)

Digitalis purpurea 'Pam's Choice' (Foxglove)

11 POPS OF COLOR

I'm getting to the back end of the perennials now but think the border could still do with a little more vertical interest and color. *Digitalis* works well for me since it can bring a sense of freedom to the planting. I want the border to feel natural, and these biennials start to travel as they see fit, which, frankly, always works better than me doing it.

A HOT, DRY BORDER

Creating a border in a garden that is naturally very dry or regularly suffers from drought can be a challenge. There are plenty of great plants, however, that can survive on surprisingly little water. I've designed this border using drought-tolerant plants that work well together across all the different layers.

Groups of plants are repeated for balance and to create a rhythmic, natural effect.

MY DESIGN

This design shows the different plant layers in action, with repetition and grouping used to bring rhythm and balance. The bulbs chosen (not shown on the plan) are an *Allium* that works particularly well in dry soil and a beautiful Turkish tulip.

1 *Lagerstroemia indica* 'Rosea' (Crape myrtle)

3 *Foeniculum vulgare* 'Purpureum' (Bronze fennel)

4 *Stipa gigantea* 'Gold Fontaene' (Golden oats)

2 *Phlomis russeliana* (Turkish sage)

7 *Eryngium agavifolium* (Agave-leaved sea holly)

5 *Panicum virgatum* 'Rehbraun' (Switch grass)

6 *Centranthus lecoqii* (Red valerian)

9 *Achillea* 'Inca Gold' (Yarrow)

8 *Limonium platyphyllum* 'Violetta' (Broad-leaved statice)

OVERHEAD PLANTING PLAN

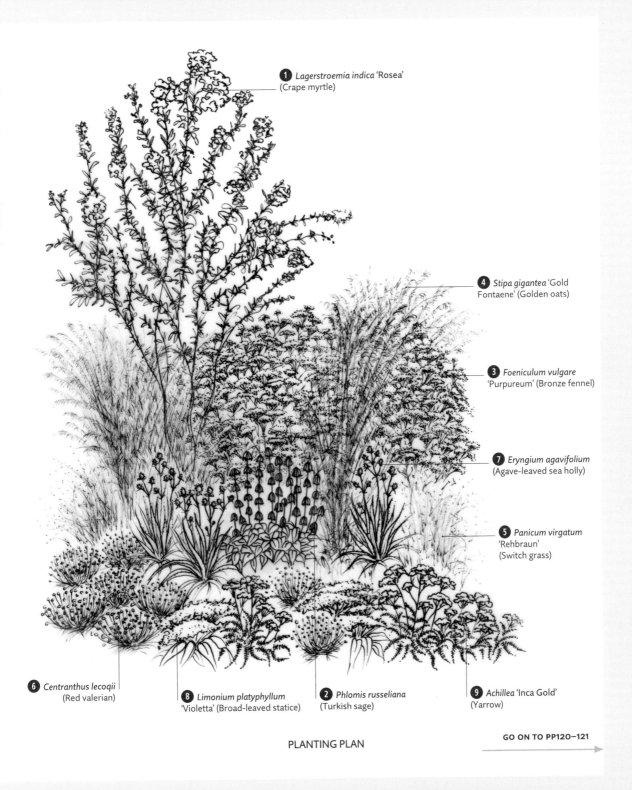

1 *Lagerstroemia indica* 'Rosea' (Crape myrtle)

4 *Stipa gigantea* 'Gold Fontaene' (Golden oats)

3 *Foeniculum vulgare* 'Purpureum' (Bronze fennel)

7 *Eryngium agavifolium* (Agave-leaved sea holly)

5 *Panicum virgatum* 'Rehbraun' (Switch grass)

6 *Centranthus lecoqii* (Red valerian)

8 *Limonium platyphyllum* 'Violetta' (Broad-leaved statice)

2 *Phlomis russeliana* (Turkish sage)

9 *Achillea* 'Inca Gold' (Yarrow)

PLANTING PLAN

GO ON TO PP120–121

HOW I WORK

The spectacular lower tree canopy sets the scene for this border. I add different shapes and heights to this framework, using color, movement, and texture to create interest and provide contrast.

① FIRST, THE CANOPY

The starting point here is the *Lagerstroemia*, which really sets the scene for the dry border. I grow this tree as a multistem, and it's a real showstopper. Here, it's the pivotal point for the rest of the border to work off. It has fantastic peeling bark, and crinkled crêpe paper–like flowers appear in late summer.

Lagerstroemia indica 'Rosea'
(Crape myrtle)

② STRONG, SIMPLE SHAPE

Next, I want to add a strong but simple shape to ground my tree, which comes in the form of *Phlomis*. It has quite a rounded habit but will spread over time. Its gray-green heart-shaped leaves have a rough texture. Stiff upright stems carry whorls of yellow-hooded flowers. This plant also looks great in winter, especially when covered with frost.

Phlomis russeliana
(Turkish sage)

Centranthus lecoqii
(Red valerian)

⑥ ADD CONTRAST

Now I have some movement and rhythm, but I want to add interest and contrast, so I look for something to feed off the shape of *Phlomis*. I add the *Centranthus* in the foreground, which has quite a strong shape. Its narrow, lance-shaped shiny green-gray leaves work well with the *Phlomis* colors, and their texture provides contrast.

⑦ BOLD ARCHITECTURE

The border is now really starting to take shape. I want it to have an architectural presence that really catches the eye. *Eryngium* is the sort of plant that stops you in your tracks. Although it's not big, it has real architectural presence with its evergreen serrated, swordlike leaves. Soft greenish-white flowers sit about 3ft (1m) above the leaves throughout the summer.

Eryngium agavifolium
(Agave-leaved sea holly)

⑧ CREATE MOVEMENT

Now I'm thinking about leaf shape to create movement, and the last two plants need to be very different. I've seen *Limonium* growing in the wild in Greece, so I know it will be happy with the other plants I've chosen. It has large rosettes of broad, leathery deep-green leaves that can change color to a bronze/red. Sprays of frothy, soft purple flowers sit on wire stems from July to August.

Limonium platyphyllum 'Violetta'
(Broad-leaved statice)

Foeniculum vulgare
'Purpureum' (Bronze fennel)

3 FORM AND HEIGHT

At the back of the border, I'm starting to think about height. I want it to be be light and feathery, which takes me to the fennel. It provides a strong form at the back. The darker ferny foliage is a great contrast to everything else. Flat-headed tiny yellow flowers appear in summer, which attract the bees. The foliage works really well with the verticals of the grasses, achieving a light, airy feel.

4 SEASONAL INTEREST

Next, I work some *Stipa* in along the back of the border, then carry the form to the foreground. I'm thinking about the longevity of the season, and the evergreen *Stipa* works well here as it provides interest into the winter. The flower stems can reach around 10ft (3m) in height, and the light plays beautifully with the golden flower head of this plant.

Stipa gigantea 'Gold Fontaene' (Golden oats)

5 CONTINUITY AND RHYTHM

Then I look at the next grass to add some continuity and rhythm, as well as provide something a little different. The *Panicum* does just that. It has a strong vertical structure and stunning fall color. It starts the year with an upright midgreen leaf that carries small clusters of purple-green spikelets, which then turn a deep rich red as fall approaches.

Panicum virgatum 'Rehbraun' (Switch grass)

9 SOFT TEXTURE

Then I wanted to add a softer texture, so I was drawn to the *Achillea*, which for me is a great little worker. 'Inca Gold' is a short, sturdy plant with silvery-green soft-textured leaves. These are topped with golden-orange flat flower heads that last well throughout the summer.

THE BULB LAYER

The bulbs I've chosen reflect the fact that I want to add a little detail that is familiar—but isn't. The *Allium* is not your normal *Allium*. It flowers a little later than many others (June–July) and has narrow midgreen leaves, which die back before the blue globe-shaped flower head appears. This goes well with so many colors and works fantastically well in gravel gardens.

Tulipa turkestanica (Species/wild tulip)

Lastly, I want to add a little interest earlier in the year, and I have a soft spot for the species tulip *turkestanica*. As its name suggests, it comes from Turkey, so I know it will be comfortable with the rest of my planting. It is simply stunning. Above the green-gray leaves sit star-shaped ivory-cream flowers with dark yellow centers. They tend to be 8–12in (20–30cm) tall and even hold a little scent.

Achillea 'Inca Gold' (Yarrow)

Allium caeruleum (Blue-flowered garlic)

DAMP GARDEN

If you find that you've got moist soil, it's best not to bother trying to fight it. I've designed this scheme using a combination of moisture-loving plants that complement each other and that will provide seasonal color and interest throughout the year.

There are lots of great plants you can grow that really don't mind having their feet damp.

MY DESIGN

With a contrast of shape, form, and color, this scheme will do well in moderately damp conditions in either sun or part shade. Underneath, but not shown on the plan, I have included a pretty *Leucojum* for spring and a *Camassia* to add midseason color and height.

1 *Betula nigra* (River birch)

4 *Molinia caerulea* subsp. *caerulea* 'Edith Dudszus' (Purple moor grass)

2 *Salix elaeagnos* subsp. *angustifolia* (Rosemary leaf willow)

6 *Persicaria amplexicaulis* (Red bistort)

3 *Cornus alba* 'Kesselringii' (Tatarian dogwood)

5 *Aruncus* 'Horatio' (Goat's beard)

8 *Trollius x cultorum* 'Alabaster' (Globeflower)

9 *Rodgersia pinnata* 'Elegans'

7 *Iris* 'Flight of Butterflies' (Siberian iris)

OVERHEAD PLANTING PLAN

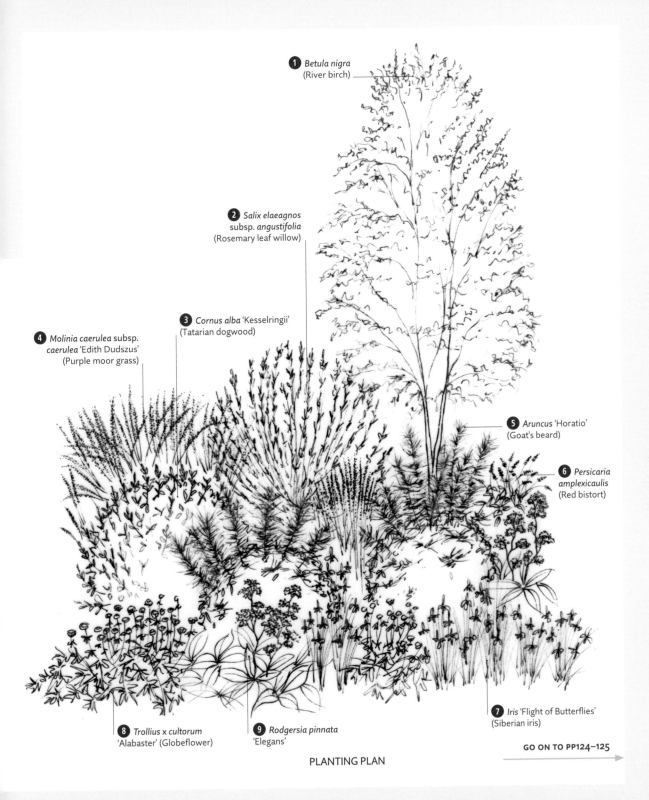

1 *Betula nigra* (River birch)

2 *Salix elaeagnos* subsp. *angustifolia* (Rosemary leaf willow)

3 *Cornus alba* 'Kesselringii' (Tatarian dogwood)

4 *Molinia caerulea* subsp. *caerulea* 'Edith Dudszus' (Purple moor grass)

5 *Aruncus* 'Horatio' (Goat's beard)

6 *Persicaria amplexicaulis* (Red bistort)

7 *Iris* 'Flight of Butterflies' (Siberian iris)

8 *Trollius* x *cultorum* 'Alabaster' (Globeflower)

9 *Rodgersia pinnata* 'Elegans'

PLANTING PLAN

GO ON TO PP124–125

HOW I WORK

I want to try to bring the tree down into the garden and counterbalance the vertical with a strong form. From here, I start to work shape to one side. Although I want good height and form, I don't want the shapes to be too heavy. Between the two shrubs, I repeat the grass from the back, which will provide good movement and contrast to the center of the border.

❶ THE CANOPY

The upper canopy starts with *Betula*, a tree that offers a lot to build the rest of the border off. It is deciduous and will grow to about 49ft (15m) high. I love to use it as a multistem since it has a light and airy feel with remarkable shaggy bark that has stunning colors.

Betula nigra (River birch)

Salix elaeagnos subsp. *angustifolia* (Rosemary leaf willow)

❷ ADD FORM

I've chosen the two shrubs not only for the form but also because they can be stalled, so I can control the growth if needed. The first is *Salix*, which is one of my favorite willows. I love its rosemary leaflike foliage, which is wonderful to plant against. It also provides a yellow fall color and has bronze-brushed stems in winter.

❻ BALANCE HEIGHT

Looking at the gap at the back on the right, I want some balance of height to the grasses and would like more color. *Persicaria* is perfect. It carries long bottlebrush-type flowers from mid-summer to early fall. The plant provides great vertical interest, and bees and other insects love it.

Persicaria amplexicaulis (Red bistort)

❼ SUMMER INTEREST

Across the front of the border, I really like to develop the summer interest with flower and leaf shape. The *Iris* not only provides color but also verticals, which start to link with the planting at the back of the border. The flowers are a purple/blue with white falling from the center and present so much detail to feed off.

Iris 'Flight of Butterflies' (Siberian iris)

❽ ADD CONTRAST

The next plant I choose is a *Trollius*, which has the most gorgeous pale yellow globe-shaped flowers. They are pure charm. These contrast wonderfully with both the *Iris* and *Aruncus*, giving a real midseason show to the border.

Trollius x cultorum 'Alabaster' (Globeflower)

3 SEASONAL INTEREST

Next to the *Salix*, I plant a *Cornus*. It can provide a good backdrop for summer interest, but in the winter, it really comes into its own and will sit well with the *Betula*. During the summer months, the leaves are oval-shaped and green before taking on a rich claret in fall. During the winter, the stems take on a dark purple color.

Cornus alba 'Kesselringii' (Tatarian dogwood)

4 COLOR AND MOVEMENT

With the first three plants providing structure, now I really want to add color and movement. To the back, I'm adding the very effective verticals of *Molina*. I like the way the flowers stand well above the neat base of foliage, which really comes into its own during fall and plays wonderfully with the light. It contrasts to great effect with the *Cornus*.

Molinia caerulea subsp. *caerulea* 'Edith Dudszus' (Purple moor grass)

5 ADD INTEREST

Next, in the middle of the border, I look at working interest in front of the two shrubs. I want the height of the next plants to be only half to three-quarters the height of the plants behind and the foliage to be different. The *Aruncus* works really well here. It has finely divided foliage and the red-toned flower stems sit well against the shrubs behind.

Aruncus 'Horatio' (Goat's beard)

9 FOLIAGE TEXTURE

The last perennial here has to have great foliage interest, and I'm drawn to *Rodgersia*. It offers real drama with its clumps of bronze foliage that later provide rich fall color. The leaves really are elegant and handsome with a beautiful texture that will provide a focal pull. Add to that its soft pink to creamy flowers, which appear in late summer.

Rodgersia pinnata 'Elegans'

THE BULB LAYER

In the bulb layer, I'm thinking color, detail, and verticals, which I want to drift through the border. I plant one main group and let the bulbs naturalize from there. The first bulb I'm drawn to is *Leucojum*, which has bell-shaped green-tipped white flowers on an arching stem. It flowers in April and May and reminds me of a snowdrop's beauty.

Leucojum aestivum 'Gravetye Giant' (Summer snowflake)

Last but not least, I want a color to work with the midseason flowering plants. One of my favorite bulbs, aside from the natives, is *Camassia*. These are tall clump-forming perennials that have a stately feel. The foliage provides a great upright with the spires of rich starlike flowers. The plant looks its best in in April and May.

Camassia leichtlinii subsp. *suksdorfii* (Caerulea Group)

REVIEW YOUR PLANTING WISH LIST

Once you have ideas from your mood board and a wish list of plants you might like to feature in your garden area, use this checklist to make sure your plants are right for both the soil and climate you have and the mood you want.

PLANT CHECKLIST

Review each plant on your wish list in turn, using the suggestions here as a starting point. The notes you make will help you work out where the plant would best be positioned, which other plants it could be grouped with, or if it needs to be dropped entirely.

❶ SITE SUITABILITY

Double-check the plant's care requirements to see whether they'll grow in your garden's conditions. Be mindful of your site's soil, climate, water table, aspect, and any microclimates (see pp28–29). If you cannot meet the plant's needs, you may need to consider dropping it from your wish list.

❷ ROOT REQUIREMENTS

Find out whether it's a thirsty plant and whether you'll need to water it more often than other plants in summer. Will you need to improve your soil or feed it regularly? How far will the roots spread? All these details will come in handy when you figure out where best to position your final selection of plants.

❸ SIZE, SHAPE, AND HABIT

Find out what the the plant's size will be when fully grown. Will it suit your space, or is there a chance it could grow too tall and dominate its surroundings? Think also about the plant's growth habit and what shape it will be when it's mature. How does it hold itself? Can it be pruned to meet your needs? Does it form a solid block, or can you see through it? These details will help you decide where best to position the plant within your garden.

Your choice of hedge will be driven by its size, maintenance needs, and seasonal appearance.

The vivid fall leaves of deciduous trees provide eye-catching seasonal interest.

5 SEASONAL COLOR AND INTEREST

Think about what interest the plant can bring to the garden as the seasons change. What does it offer in terms of foliage, flowers, berries, seed heads, bark, or winter skeletons? How long is its flowering season, and does it flower repeatedly? Take into account not only the color of the flowers but also that of the leaves, stems, or trunks, berries, and seed heads, and also how they might change over the year. Is it evergreen or deciduous? How does it look when it dies back? Use this information to plan for a garden that will interest you throughout the year.

THE FINAL EDIT

By the end of this process, you should have a shortlist of plants you are confident will thrive in your garden area and look good in it. It might take a while to gather all of the information, but it will be worth it when you come to create your final design.

6 MAINTENANCE

Think about the practicalities. What about pruning and training, cutting back, deadheading, and dividing? How much attention will it need over a 12-month period? Will it self-seed? Better to find these facts out now than later on, after you've positioned a plant in an awkward spot where you're unable to reach it with the pruning shears.

7 SHADE

Don't forget about the impact the plant might have on its neighbors—or your own, for that matter. Will it cast shade and, if so, how dense will the shade be? Will it cast a shadow over the garden next door? If so, you may need to take care where you position your plant, or keep it separate from other, more sun-loving plants.

Delphiniums offer towering spires of flowers.

4 FLOWER SHAPES

The shape of any flower will have an impact on the look of your overall design. Does the plant form flowers in the shape of spires, umbels, daisies, buttons, globes, plumes, goblets, or trumpets? Make a note so you know what you can expect when your plant comes into flower. You should be looking to have plenty of contrasting flower shapes on your final list.

Make sure you have lots of contrast in the leaf shapes you choose.

Ask ADAM

It's so easy to get carried away and buy plants you've fallen in love with. However, to make sure you don't end up with a muddle of one of this and two of that, do your research and think of plants in groups rather than as individuals. Here are some questions I'm often asked.

Q DO I HAVE TO **REPLANT TULIPS** EVERY YEAR?

In theory, tulips should reflower every spring. However, apart from species tulips, most are best replaced every year since, for various reasons, you can't rely on them to reflower year after year. I always recommend planting tulips in containers near the house so you get a hit of color every day. Later, at the end of the season, they are easy to remove and replace.

Q WHERE DO **CLIMBERS** FIT INTO THE **LAYERS?**

If you approach planting by looking at layers (see pp92–93), then you're off to a great start. However, plants such as climbers and hedging don't always fall neatly into one of these categories, so I have included them in a separate list (see pp104–105 and pp108–109). Most climbers I would class in the shrubs layer. Clothing your boundaries can really help affect the atmosphere of your garden in a positive way. If you're considering planting a climber, it's also worth checking whether it is self-clinging or whether you'll need to install wires or a trellis to support it.

QUICK FIX

HOW TO ADD SOME **QUICK COLOR** TO YOUR GARDEN

If you want a garden area full of plants as soon as possible, annuals are a good option. They grow, set seed, and die in the space of a year, and many can quickly grow to a substantial size. You can give annuals a head start by sowing them early on in the year under cover, but if you don't have the space or time, simply sow the seeds in the ground in spring where you want them to grow.

*If you include plants from **each of the layers**, you will always create a **good sense of balance** in your design.*

Q HOW OFTEN SHOULD I **WATER** PLANTS?

It's best to keep seedlings and anything newly planted fairly moist and never let the roots dry out completely. In hot weather, pots often need a good dousing every day. With garden plants, much depends on how well a particular plant is adapted to cope with drought. Look out for signs such as drooping or shriveled leaves. You can test the soil by sticking your finger in it. If the soil is dry, the plant needs watering. It's better to give plants a good soak rather than watering little and often. Remember that mulching can help retain moisture.

Keep wildlife in mind when you choose your plants. Try to provide **a good mixed habitat and supply of food** to encourage birds, bees, and butterflies into your garden.

Q SHOULD I BUY A **CONTAINER** OR **BARE-ROOT** TREE?

The deciding factors here are timing, cost, and availability. Container-grown and root balled trees and shrubs can be planted pretty much any time of year, although it's harder for them to establish in summer. Bare-root specimens (which are field grown but stripped of earth) are a cheaper option but can be planted only in fall and winter. You are best buying stock.

KEY *knowledge*

○ Choose plants that will thrive in your soil and climate. There's no point in growing things that won't be happy.

○ Try not to buy plants on an ad hoc basis, but think ahead about exactly where things need to be planted and how they will work together.

○ The most important part of planting that you need to get right is your structural planting of trees and shrubs.

○ If you include plants from each of the layers—upper canopy of trees, lower canopy of trees, shrubs, perennials, and bulbs—you will always create a good sense of balance in your design.

○ When planting, allow enough space for plants to reach their full size.

DESIGN

BRING IT ALL
TOGETHER

> *This is the exciting part when all your fact-finding and research come together into a final design.*

INTRODUCTION

Now the fun starts, when all your research comes into play. I'd encourage you to experiment with different options before deciding on your final layout and not rush to get to the end too quickly. This will help you create a garden that will really work for your needs and in which you'll love spending time.

There isn't one fixed way of designing gardens. You just have to find an approach that suits you. In this chapter, I've set out one way that you could think about the process, and for each stage of it, I've included a garden I designed, so you can see how my mind was working as I developed my ideas.

Whether you are working on your whole garden area or just revamping it one border at a time, it's easier to break the process into smaller steps, as the whole thing can become overwhelming. Garden design is a big subject, so take your time, explore different ideas, and think things through as much as possible before putting anything in place.

Also, don't get hung up on specifics or think that a great garden has to be all about money. If the garden area is designed well, you can build it in stages and be really clever with your use of materials, or plan on upgrading things in the future as and when finances allow.

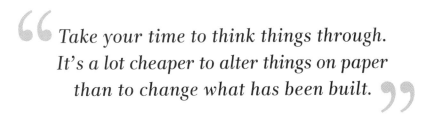

Take your time to think things through. It's a lot cheaper to alter things on paper than to change what has been built.

STEP 1

START THE PROCESS

Don't think that you need to be great at drawing for this. You don't. At first, it's all about playing with ideas to explore how to achieve a design that is both beautiful and works with your day-to-day life. Only in Step 4 will you start to think about details and costs.

❶ IN THE BEGINNING

Earlier in this chapter, I showed you how to draw up an accurate plan of your garden area (see below). It will help you understand the space you have. Annotate it with fixed elements already in place (such as trees and sheds), and make sure it shows areas of sunshine and shade.

❷ LET THE IDEAS FLOW

You draw the plan only once; then, for each different design sketch you do, you simply overlay the plan with tracing paper and hold it in place with masking tape. On your first sheet of tracing paper, then, begin to sketch out roughly some basic shapes and ideas (see opposite).

❸ FIND THE VIEWS

Make sure you set your plan up as if you are looking out from the house, as that is your main connection with the garden area on a day-to-day basis. Are there some strong sight lines from the house or the terrace?

WORKING ON A SCALE PLAN

This is your springboard for the design process. Below I have started to note on my plan some of the key points about the space I'm going to work on.

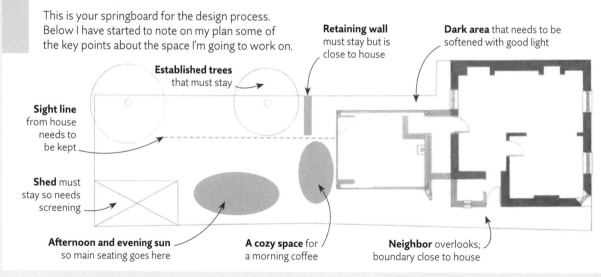

Retaining wall must stay but is close to house

Dark area that needs to be softened with good light

Established trees that must stay

Sight line from house needs to be kept

Shed must stay so needs screening

Afternoon and evening sun so main seating goes here

A cozy space for a morning coffee

Neighbor overlooks; boundary close to house

HOW I WORK

Overall, I want the garden to have a relaxed atmosphere and to feel "comfortable"—as though it has been there for a long time. I start by sketching out various different ideas, playing with the space. It's not about detail; it's just about understanding what I've got.

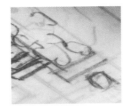

At this point, I'm imagining the space and moving my way around, in my mind's eye, to figure out how I want people to use the garden area.

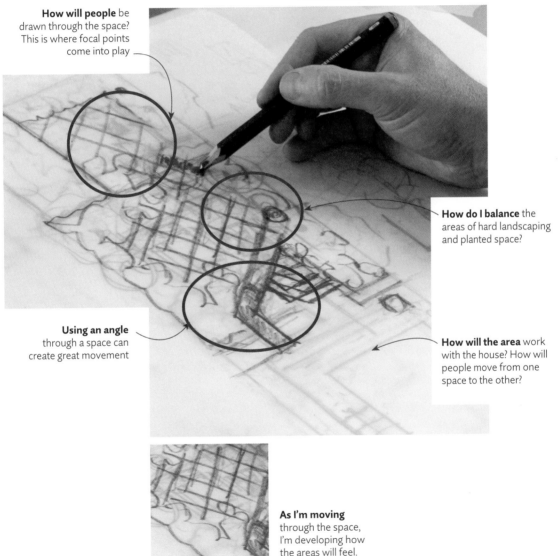

How will people be drawn through the space? This is where focal points come into play

How do I balance the areas of hard landscaping and planted space?

Using an angle through a space can create great movement

How will the area work with the house? How will people move from one space to the other?

As I'm moving through the space, I'm developing how the areas will feel.

GO ON TO STEP 2

STEP 2
EXPLORE OPTIONS

Think about your key areas: how you can link them together, how people might move through the space, and what atmosphere you want. Don't rush this stage, just keep exploring ideas. How will people feel in the different spaces as they sit? What view will they have?

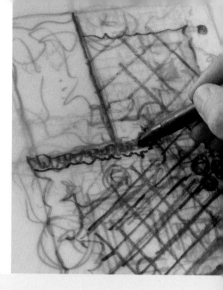

1 TRY DIFFERENT SHAPES

Experiment with different shapes for the key areas in the space, using a fresh sheet of tracing paper for each version. Try a variety of shapes: strong, fixed shapes; soft, flowing shapes; squares, circles, or ovals, either distinct or overlapping. Vary the proportions of your ovals and rectangles, for example, to see how they occupy the space and sit with each other. See what seems to work well shape-wise in your space (see opposite). Make sure you do lots of options. This will create not only new, usable spaces but also movement.

2 SOMEWHERE TO SIT

Revisit your earlier research on the number of people who will use the sitting space to focus in on any terracing you want to have (see also pp66–67). Check on your plan where the sunniest spots in your garden area lie. Is this where you want to sit people? As well as thinking practically, it's important to think about how you want the space to feel—like an extension of your home, for example, or like a woodland glade?

3 KEEP PLAYING

Explore multiple options on fresh pieces of tracing paper. Remember you can guide and manipulate the way people move through a space, using focal interest to create movement, and the materials you choose for pathways, for example, to control how they travel (see pp66–67). Using focal points and planting can draw people around a space, while hedging, screening, and framing devices can build up elements of excitement, anticipation, or calmness along the way.

> *Just like the rooms inside a home, gardens are made up of various elements that, combined together, give a sense of style and space.*

HOW I WORK

As I start experimenting with ideas, I'm just trying to keep things simple and think about the bigger ideas than the specific shapes.

OPTION A
Using angles can create movement and open up an area, but watch out that it doesn't generate awkward spaces.

OPTION B
Wide stairs up from the house can open the link to the next space, while built-in seating helps get more space close to the house.

OPTION C
Dividing the space to create "rooms" can help when you want zones for different functions and different atmosphere, but it can also make an area feel too small.

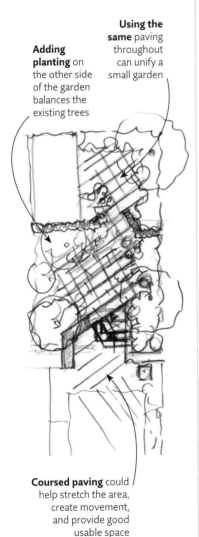

Using the same paving throughout can unify a small garden

Adding planting on the other side of the garden balances the existing trees

Coursed paving could help stretch the area, create movement, and provide good usable space

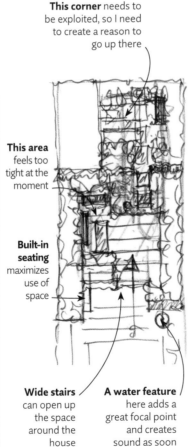

This corner needs to be exploited, so I need to create a reason to go up there

This area feels too tight at the moment

Built-in seating maximizes use of space

Wide stairs can open up the space around the house

A water feature here adds a great focal point and creates sound as soon as people step out of the house

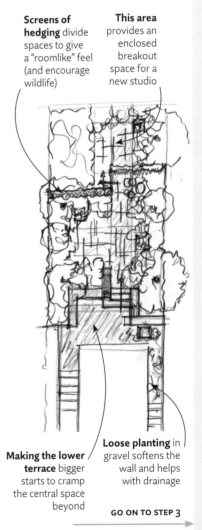

Screens of hedging divide spaces to give a "roomlike" feel (and encourage wildlife)

This area provides an enclosed breakout space for a new studio

Making the lower terrace bigger starts to cramp the central space beyond

Loose planting in gravel softens the wall and helps with drainage

GO ON TO STEP 3 →

STEP 3

FIND THE RIGHT BALANCE

The best design is normally the one that feels the most balanced—one that meets your requirements for the proportion of space allocated to planting and that given to general usage. Keep playing until you're happy that the garden works, both practically and aesthetically.

❶ GO OUTSIDE

Being in the space is often the best way of understanding what will and won't work, so keep going out into the garden area with your ideas. You can always mark shapes out with canes or lines to see how they work in the physical space. Stand in the garden area with your plans, trying different options for what goes where and mapping out different shapes until it feels right. Imagine how people will move around the space and what they'll be looking at.

❷ THINK IN 3-D

Keep checking views—both good and bad—and whether you have exploited them or blocked them to your satisfaction. Think about how your different options might tie in to the architecture of your home. Check that the atmosphere you are after has been achieved. The main thing is to stay open-minded. When you've figured out the best way to divide up the space, go back and fine-tune it (see opposite).

❸ FIND YOUR FIT

Deep down, most of us instinctively know when a place feels comfortable or not. I often give the example of moving into a new house and arranging your sofa and chairs. Maybe you'll begin by putting them facing the focal point of the fireplace or the TV. You move them into position, and it feels all horrible. So you go for a cup of coffee, then move them around again; and with just a few adjustments, it all feels just right—comfortable and workable. For me, that is the feeling you are looking for from your garden area.

THE RIGHT BALANCE

We all see space differently, but a successful garden area for me has a good balance of planting and usable space. You might want to get another piece of tracing paper, overlay it over your potential design, then color the planting spaces in and leave the usable spaces blank. Do you feel that they are balanced against one another? This circle/square example just gives you an idea of what you're looking for.

A dominant square dwarfs the circle within it, making it feel uncomfortable.

More equal proportions between the circle and square create a more balanced feel.

A dominant circle overpowers the square and leaves awkward space around the sides.

HOW I WORK

As my space comes together, I keep asking myself what patterns of movement I want. Do I want people to pause as they move through the space? Am I providing good views? How can I start to add the detail that makes it a personal space? How do I create that atmosphere?

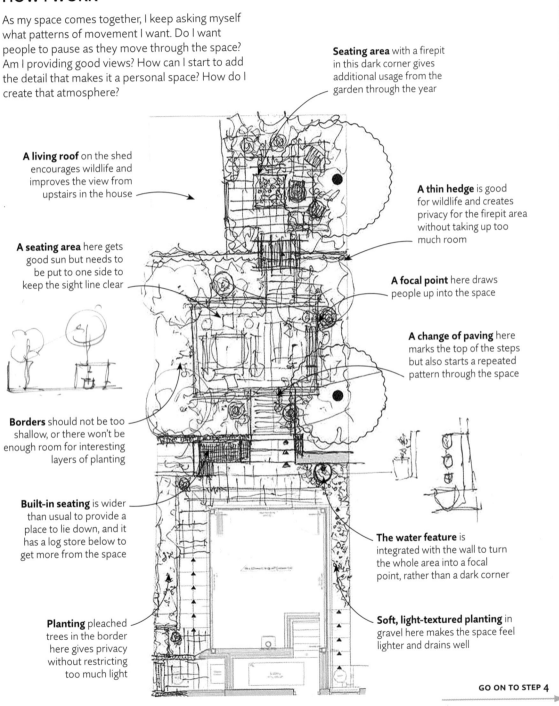

Seating area with a firepit in this dark corner gives additional usage from the garden through the year

A living roof on the shed encourages wildlife and improves the view from upstairs in the house

A thin hedge is good for wildlife and creates privacy for the firepit area without taking up too much room

A seating area here gets good sun but needs to be put to one side to keep the sight line clear

A focal point here draws people up into the space

A change of paving here marks the top of the steps but also starts a repeated pattern through the space

Borders should not be too shallow, or there won't be enough room for interesting layers of planting

Built-in seating is wider than usual to provide a place to lie down, and it has a log store below to get more from the space

The water feature is integrated with the wall to turn the whole area into a focal point, rather than a dark corner

Planting pleached trees in the border here gives privacy without restricting too much light

Soft, light-textured planting in gravel here makes the space feel lighter and drains well

GO ON TO STEP 4

STEP 4

ADD SOME DETAIL

You are now at the point where your design works on a practical level, and the proportions and aesthetic feel right. It's time to use your style mood boards to fill in the details around your choices for the hard landscaping, such as pathways and focal points.

> *Your choice of materials and details can ground the garden area in its location and make it a personal and unique space.*

1 CHOOSE MATERIALS

Start thinking about the areas of hard landscaping that will have the biggest visual impact—terracing, steps, and paths, perhaps. First, look at your terrace or seating area. How do you want to link it to the architecture of the house? What materials did you think would work best when you explored the options earlier in this chapter (see pp60–63)?

2 CONNECT SPACES

Now think about texture and materials to plan the paths (see pp64–67). Remember how you want people to feel in the space and how you want them to move around. Do you want a slow, meandering path? Do you have areas of heavy traffic that suits a hard-wearing, practical surface? Using different materials can delineate areas for different purposes.

3 DEVELOP YOUR STYLE

Repeat this process for the other hard landscaping elements in your garden area: lighting; pergolas; focal points; and other features, such as sculptures or water channels. Draw on your earlier research on hard landscaping to work in your ideas and style preferences. Try to build a sense of cohesion so that the space has its own sense of place. Don't worry about fine-tuning elements such as furniture (unless it is built in) until you have finished the garden area.

The stone of the walls has been picked up in the paving to bring a lovely cohesion to the space.

The step from the path to the terrace creates a sense of arrival and prevents the gravel from getting on the terrace.

Even the smallest feature in the garden area can provide a point of interest and enjoyment.

The final details in your garden area should reflect your individual character and your personal tastes.

4 CHECK YOUR BUDGET

Earlier, while you were designing the space, thoughts of money shouldn't have been interfering with your ideas—if you organize your space properly, the materials are of secondary importance. At this point, however, you do need to check your choices against your budget. Need a large terrace but can't afford the expensive stone you've had your eye on? There's no point building it any smaller, so use a cheaper option this time around—you can always upgrade it later!

5 BRING IN PLANTS

Before I finalize my layout plan, I do start thinking about plants, but only at the structural level. I won't know which plants I want, just whether I need to frame a view, perhaps, or screen a neighbor, or maybe add a little rhythm through the space. Do I need a tree to create a secluded seating area? Do I need to add a tree to balance with existing trees? This is the point at which these trees become part of the structure of your garden area.

GO ON TO STEP 5

STEP 5

FINALIZE YOUR PLAN

Having a final plan will help keep you focused and mean you can price your build, although your plan may evolve as you start to landscape. It doesn't have to be a work of art; it just has to give you a scaled understanding of your new spaces.

❶ REVISIT AND REVIEW

Step back and review where you've got to with your design. How does the space feel? Is it looking too busy? Perhaps you need to pare down the different elements into a more unified approach. Is it lacking something? Maybe you need something bold to shake it up a bit. Is it practical where you need it to be practical? How does it feel, and will it lead people around in the way you want them to experience the space? Does it deliver the atmosphere you want?

❷ DRAW YOUR PLAN

Draw your final design accurately to scale to use as your master plan, adding in all the details. Don't be afraid to make any final adjustments if you need to—it's easier to alter things on paper before the build starts than to change something that has already been built.

❸ YOUR FINAL DESIGN

At the end of the process, you should have a scale drawing that reflects all your design and structural planting decisions. From this, you can accurately figure out quantities for the materials you need and do your final pricings.

❹ NOW TO THE PLANTS

You have your garden layout and materials locked down now, so you can finally start to think about the planting spaces. So far, it's just been about the structural elements, such as the bigger trees and shapes. Now it's time to focus on planting in more detail.

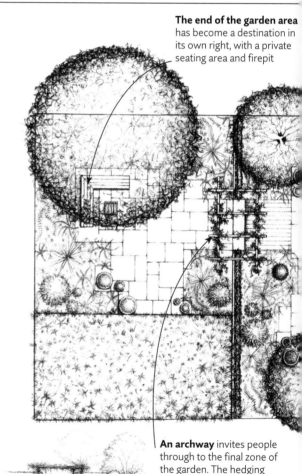

The end of the garden area has become a destination in its own right, with a private seating area and firepit

An archway invites people through to the final zone of the garden. The hedging keeps the shed area screened off and offers sanctuary to wildlife

Hedging and archway

HOW I WORK

I'll work my final design up as an accurate overhead plan and also as little 3-D sketches to visualize how the spaces will look and how features will work. Yours do not have to be beautiful; they are more just to help you understand how things will look.

Your plan doesn't have to be a work of art; it just has to be accurate and give you a clear reference to build from.

Directly outside the house is an integrated seating area, with planting that frames the view up the garden area and a water feature that can be enjoyed not only from the house but also from the seating area

Integrated seating area and water feature

The main seating area sits to one side to enjoy maximum sunshine and avoids interrupting the sight line from the house

A square of setts (bricks) provides detail and a lovely pause point

A slim row of pleached trees sits along this wall to screen off the neighboring house without blocking light

GO ON TO STEP 6

STEP 6

BUILD YOUR LAYERS

Once you have your final garden layout, start thinking about planting. Just look at plant shapes and spaces for each layer first. Using colors to represent different areas or shapes, build up layers that flow and connect spaces. Try several versions until you're happy with the result.

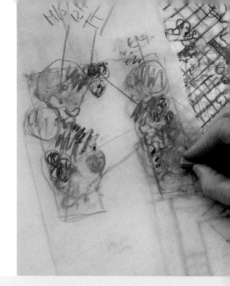

❶ BUILD YOUR LAYERS

On fresh tracing paper overlaid onto your design, add in your structural framework of trees and shrubs, using color to denote their form "type" (see box below). Make sure you have marked your trees in the right place, then start to add your shrubs and perennials. Build up your design in terms of shapes and forms, using colors to represent different shapes and types of plant ("tall and thin," for instance). You may want to revisit my border designs for ideas (see pp110–25).

❷ CREATE GROUPS

As you build your layers, think in groups, not in ones or twos. Plants in nature grow in clusters, with a few single plants setting themselves a short distance away, too. If you mimic this, you'll get a sense of flow and rhythm to your planting. Think back to how the layers of natural woodland work and follow nature's lead (see pp92–93). I'm very driven by nature. I love the way groups of plants overlay each other and the contrasts you find naturally between foliage shapes and textures.

❸ RHYTHM AND FLOW

Build in repetition from one border to another to help the different parts of your garden area connect and relate to each other. Using the same plant or plants with a similar shape can create a strong sense of rhythm and flow, so repeat a few plants in different areas as a way of linking them. Likewise, a few strong individual plants can introduce drama and create a focal pull. Keep in mind, too, your mood board on the style of planting you want to create (see pp90–91).

THINK ABOUT FORM

Plants offer different shapes, or forms, such as these examples, that you can exploit for contrast, rhythm, and repetition.

Compact, dense, rounded mound with small leaves

Low-growing mound with flowering spires

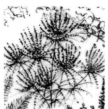

Solid and bushy with a striking structure

Tall bush with oblong leaves and open habit

Light, feathery fronds with lots of movement

HOW I WORK

For this stage, I work rapidly, focusing only on layers, flow, repetition, and shapes, rather than specific plants. Try to avoid getting bogged down in detail. What I'm trying to do is figure out how I want the garden to feel and how I imagine being in the space.

Detail is not important at this stage; what is key is getting rough heights, forms, and scale in place

Balance existing trees by adding structural plants on the other side of the garden

Building up the layers, starting at the top and working down, keeps a sense of flow and balance

Adding drama with a few keynote plants will act as natural focal points and stop people in their tracks

Bend the rule that says always put big plants at the back of a border; if you want to put something big at the front, do—just not something too heavy

Repeating shapes helps move the eye through the space and gives a sense of flow to the planting

Using color helps me visualize the planting and understand how the layers are working in different areas. That does not mean the plants will be those colors—it just helps me understand

Don't rush: I'll revisit my ideas five or six times, reviewing and detailing as I go. Each time, I'll add more detail and drill down into different elements, such as contrasting leaves, flower shapes, and seasonal interest (see pp146–147).

Grouping perennial plants into clusters gives a more natural effect, mimicking nature's way of growing in groups

Creating rhythm by repeating plants regularly throughout the garden links areas together and builds cohesion

GO ON TO STEP 7

STEP 7

PLOT YOUR PLANTS

Once you're roughly happy with the layers, rhythm, and flow of your plan, you can start to go through your plant list and pencil in specific plants that would fit the bill, keeping in mind the atmosphere you want to create.

❶ USE YOUR LISTS

Go through your plant lists for each layer to see what matches the forms you've planned. Start allocating specific plants to the shapes you want. Factor in what they will bring to the party in terms of seasonal interest (such as blossom, spring leaves, berries, fall color) and anything else that's particularly good about them (bark, fruits, and so on).

❷ GET COLORFUL

Now refine and adjust your plant choices to factor in color and scent. For me, plant shapes, texture, and forms are more important than color, because color comes and goes in the garden, so it's not the main driving force. Don't get me wrong—I do love flowers—but color can also come from leaves, stems, and berries. As for scent, plan to put perfumed plants where you're going to pass them or sit by them, adding a bit of repetition by drifting the same plants through a border or placing single beats or punctuation marks.

❸ THINK ABOUT CONTRAST

If you can't make up your mind what plants might work together, just think about contrast. Do the plants have contrasting sizes, shapes, flowers, leaf forms, texture, and color? Contrast will always make your planting more dynamic. Planting design is a vast subject, but to begin with, just make sure you cover the essential elements here and enjoy the process!

Blue blends with so many other colors, it can be very useful fallback if you're nervous about color and need a safe choice.

> *It's not just about the beautiful details of individual plants but also about how they all work together season by season.*

HOW I WORK

At this point, I start to think about the detail, going through my plant lists for the project. As I'm penciling in specific plants that I think will fit the bill, I keep in mind contrasts of leaf, texture, and flower. Even at this point, I am still asking questions about seasonal interest and how plants will sit together. I might review this five or six times before I finalize my plan.

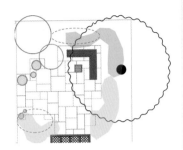

Drifts and bulbs can be plotted on a separate overlay of tracing paper, showing where you might want a "drift" of plants or a bed of bulbs underneath the main planting.

Think about color, not just in terms of the flowers, but also the bark, foliage, and berries

Repeat a few plants in different areas as a way of linking them, creating rhythm and cohesion

Contrast is key; add interest with contrasting shapes, textures, or forms

Plan for interest every season, thinking about where your points of interest are

Build drama with statement plants as focal points

Remember to think about atmosphere—how you want the planting to feel

Think about how groups will work throughout the year. Make sure that, when the plants are dormant, there won't be a big hole in your border.

Scent the air with perfumed plants in areas you want to meander past or relax in

Put winter interest near the house as you may be less inclined to walk down the garden in winter

Think about what you will be looking at from inside the house and where you will sit throughout the year

Ask ADAM

I don't think there is only one way to design a garden. I've shown you an approach that keeps it as simple as possible, but you may develop your own tools along the way, too. The important thing is not to be overwhelmed; take your time and develop your ideas bit by bit.

Q CAN I PLANT A FEW THINGS AT A TIME?

Yes, of course, you can. This is very much how I garden at home, but for me, it's still important to have an overall planting plan, even if it does evolve over the years. If you are going to plant over time, it's best to concentrate initially on your trees and shrubs, as they will be slower to reach their full potential, whereas your herbaceous plants will reach theirs in a couple of seasons.

Q WHAT IF MY BUDGET ISN'T ENOUGH?

Don't let money influence your design. The priority is to organize your space properly; the materials are of secondary importance. If you need a big patio but can't afford the stone you want, don't compromise by making your patio smaller. Choose a cheaper option such as gravel instead, and then upgrade later. Or break down the build and do it a section at a time over a few years.

A common mistake is to make **borders too narrow** with not enough room for plants to reach their full potential. If in doubt always be more generous with the space.

Q STRAIGHT LINES OR CURVY EDGES?

One isn't better than the other. There's no right or wrong answer; just follow your instincts and choose the look that feels best to you. Working through the Design chapter will give you a good sense of your own style preferences, and your garden design needs to reflect your individual choices, not follow someone else's style. Just remember, most of the lines will probably become blurred anyway as the plants spill over and soften the edges.

> *Some people think*
> *there's a **fixed way of***
> ***designing gardens**, but in*
> *reality there isn't.*

WHAT ARE RIBBONS AND **BLOCKS**?

You may hear designers using terms such as ribbons, blocks, and intermingling. They are just ways of describing planting techniques. Ribbon planting means you have wiggly line or thread of a certain plant running through the border. It draws the eye along and helps create a sort of unity.

Block planting is when you plant groups of plants in clumps next to each other.

Intermingling is inspired by the way plants grow in nature, where different plants are all densely meshed together.

QUICK **FIX**

MIMIC NATURE IN **YOUR** GARDEN

Some people buy plants in odd numbers—threes, fives, sevens, etc. That was the way I was taught, too, but to be honest, I don't really do that any more.

Besides, even if you do plant in odd numbers, a year or so later, it won't look like that, as things tend to spread around or fail. Now I'm more influenced by how things work in nature. I look at leaf contrast, form, rhythm, and texture. I find building little groups and offsets is a lovely way to put your garden together.

KEY *insights*

○ I have a rule that if you try a plant three times and it doesn't flourish, forget it. Everything should really earn its place in your garden, and if one plant can't survive, then find something that will.

○ If you are trying to get privacy in your garden, don't be drawn right away into planting a row of trees at the end of the garden; it can make the whole space feel shorter. It can be better to put smaller trees around a terrace to create a cozy area and screen your view that way.

○ Don't fight nature. As much as you might like a lawn, for instance, if your garden is too shady or small, use the space differently instead. Maybe you could put in gravel with planting and another seating area. Sometimes it's just best to work with what you have.

BUILD YOUR GARDEN

PLANT YOUR GARDEN

BUILD

BUILD

BUILD YOUR GARDEN

"Good ground prep is crucial to any successful build—don't even think about skipping it."

INTRODUCTION

For me, building a garden is one of the most satisfying things you can do. It's really not that difficult, and if you give it a go you'll have a chance to make a tangible impact in your space by adding a terrace, fence, or structure that can be used and enjoyed for years to come.

The step-by-step processes in this section cover a selection of hard landscaping techniques—in other words, permanent features like walls, pathways, and pergolas. Whatever you build, always think about its function first and style second, and keep your use of different materials to a minimum.

Building your own garden can be hugely rewarding and can also help keep costs down. Whatever you choose to build, always double-check that you're happy with your design before you start work. It's a lot easier to alter things on paper than to change them when you've already bought the materials—or even started building—although it doesn't mean you can't tweak things if you need to.

Even if you decide to employ a landscaper, the information in this section will help you understand what processes are involved in hard landscape work and what you're actually paying for.

> " *It's always, always worth checking your measurements and levels one last time before you start work on the ground and as you go along.* "

BASIC TOOLS FOR HARD LANDSCAPING

Certain tools are essential for building the solid structures in your garden. If you don't have some of these tools, you may be able to borrow from friends or neighbors, or rent them if necessary.

8. Rubber mallet

9. Carpenter's square

1. Bucket

2. Cordless drill

3. Line blocks

4. Gauging trowel

18. Line and pins

5. Shovel

7. Wire brush

20. Brick bolster

6. Spirit level

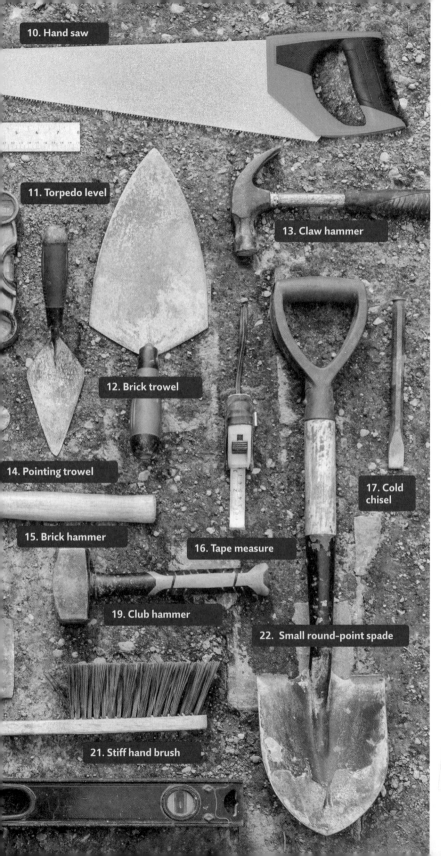

10. Hand saw

11. Torpedo level

13. Claw hammer

12. Brick trowel

14. Pointing trowel

15. Brick hammer

16. Tape measure

17. Cold chisel

19. Club hammer

22. Small round-point spade

21. Stiff hand brush

MISCELLANEOUS TOOLS
1. Bucket for measuring and carrying materials and water **2. Cordless drill** for drilling holes in wood and masonry (it's useful to have a range of drill bits, including a screwdriver)

BRICKLAYING TOOLS
3. Line blocks used with a string line to set up straight lines when laying bricks **12. Brick trowel** for laying bricks **14. Pointing trowel** for pointing mortar

MEASURING AND LEVELING TOOLS
4. Gauging trowel for applying mortar to tight areas **6. Spirit level** for checking and setting levels over large areas **9. Carpenter's square** and builder's square for marking wood and checking corners **11. Torpedo level** for setting levels in smaller areas **16. Tape measure** **18. Line and pins** for marking out areas to build off and for setting levels

SPADES AND SHOVELS
5. Shovel for use with aggregates, sand, and cement **22. Small round-point spade** for digging in confined spaces

BRUSHES
7. Wire brush for cleaning stone and brickwork **21. Stiff hand brush** for general cleaning up

HAMMERS AND MALLETS
8. Rubber mallet for tamping paving slabs and bricks down **13. Claw hammer** has a claw for removing nails **15. Brick hammer** has a sharp chisel-shaped head for chipping off edges of bricks and stone **19. Club hammer** for heavy-duty building work

SAWS AND CHISELS
10. Hand saw for cutting wood **17. Cold chisel** for chipping out small, precise details on concrete **20. Brick bolster** for cutting bricks, concrete blocks, and paving

OTHER TOOLS (not shown)
Spray line for marking out lines
Mixing tray for mixing cement
Rake for moving base rock and gravel

SAFETY FIRST

Always wear **safety goggles, heavy-duty gloves,** and **steel-toe boots** when using these tools to protect yourself from injury, and make sure a **first aid kit** is available.

PREPARING THE SITE

A key part of the work of any horizontal build—be it a gravel path or a terrace—is in preparing the site and creating a good base. Preparing the ground properly and measuring your site accurately will ensure that whatever goes on top has a strong foundation, has the right dimensions, and is in the right place.

If you're extending a terrace or updating materials, remove the current paving and investigate the foundations below. To discover whether there is preexisting base rock, dig a small hole to a depth of 2–4in (5–10cm). If you find that the base is even and solid, you can build from that and skip ahead to the "How to" sections. Or you may just need to top up the layer of base rock and then compact it down.

For paved areas next to the house, it's a good idea to make sure the paving will lie at least 6in (15cm) below the dampproof course membrane so rain cannot bridge this line. Remember that all paving should slope away from the house.

MARKING OUT

A key part of the job is transferring your scale drawing onto the ground area. Take your time. Measure and mark out carefully, put in lines and pins (to work to later), and check your corners. I usually make my working area about 8in (20cm) bigger all around to give myself a little flexibility and room to work. And always check and recheck your measurements *and* levels before you start the digging and as you go along.

Use spray line to help transfer the horizontal dimensions of your project from the scale plan to the ground itself.

Put in a series of lines and pins to mark out your site and set your final levels. Set the pins outside the working area.

THE LAYERS OF GOOD FOUNDATIONS

A well-prepared site consists of a layer of compacted ground on top of the subsoil, a layer of base rock on top of that, and the sand or mortar bed. For paving and paths, you will need to allow about 6–8in (15–20cm) for these layers. Use pegs to ensure that the layers are level and sit at the right height for your structure (see below).

> *As with many building projects, the work you do in preparing the ground itself and creating a good base is the most important work you'll do.*

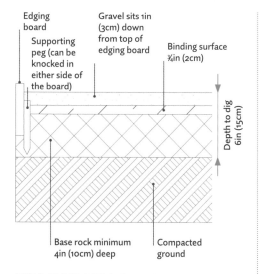

Edging board

Supporting peg (can be knocked in either side of the board)

Gravel sits 1in (3cm) down from top of edging board

Binding surface ¾in (2cm)

Depth to dig 6in (15cm)

Base rock minimum 4in (10cm) deep

Compacted ground

FOR A GRAVEL PATH WITH EDGING BOARDS

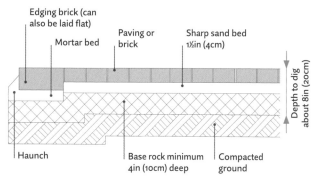

Edging brick (can also be laid flat)

Mortar bed

Paving or brick

Sharp sand bed 1½in (4cm)

Depth to dig about 8in (20cm)

Haunch

Base rock minimum 4in (10cm) deep

Compacted ground

FOR BRICK PAVING

Check that right angles are truly 90° using a carpenter's square or the 3-4-5 triangle rule (see p245).

SETTING LEVELS

In addition to marking out horizontally across the site, you need to consider the vertical dimensions—how deep to dig and how to mark the different layers to follow: base rock, mortar, paving, gravel, or brick. You start by knocking in pegs at the perimeter of your site and checking these are at the final level you want your finished surface. Remember to factor in the slope you'll need for runoff (see p242). Then you add more pegs within the site area as you dig it out and make sure these tally across the site.

Using wooden pegs helps you to construct and check the level for the finished surface all across the site.

CREATING A GOOD BASE

Once you've marked out the site, it's time to put a spade in the ground. Creating a good base is probably the most important thing you will do when building your garden. A good foundation is a solid, level, and compacted layer of base rock. Use plenty of reference pegs across your working area to help you keep track of the levels.

YOU WILL NEED

Use your plan and the instructions on pages 245–247 to calculate quantities for

○ base rock

Tools

○ tape measure

○ line and pins or spray line

○ club hammer

○ carpenter's square or 3-4-5 triangle

○ spirit levels, large and small

○ spade

○ shovel

○ mini digger (optional)

○ sod cutter (optional)

○ wheelbarrow

○ wooden pegs for marking levels (have plenty of these)

○ rake

○ plate compactor or tamper

○ iron bar (optional)

1 Mark out the area with line and pins and check your angles (see pp242–244). Then, dig down about 6–8in (15–20cm), depending on the project you are doing, to allow for the foundation layers and the thickness of your chosen surface, such as paving, bricks, or gravel. Dig out the area by hand with a spade or use a mini digger.

2 Once you're at roughly the right depth, tread the ground across the marked-out area so it's firm and even. Any soft patches will reveal themselves as your heel sinks in. If it does, dig out further until you reach a firm surface.

6 Make sure that all the pegs across the site are at the correct level. If this is your first build, use a lot more pegs than I've done here, to give yourself plenty of reference points. Once they're done, check them all again.

3 Reset your pegs or at least check that the levels are okay. Set the first peg at the perimeter of the site, level with the line of the finished surface. Set a second peg, and use a level to make sure the two are level; repeat with more pegs across the site.

4 Measure down from the finished surface level to mark on the pegs the levels for base rock and any other layers. Levels are really important—keep checking as you go. Once you have done this a few times, you may not have to keep checking as you do get a sense of levels.

7 Spread base rock to a depth of no less than 4in (10cm), using the marks on your pegs as a guide.

5 Most terraces and paths need a slight slope so that rain can drain easily (see p242). Allow for such a slope by setting up a second series of pegs to show how the level changes across the entire site. The slope on your paving may vary depending on the material—check with your supplier.

8 Knock and rake level, then tamp the base rock with a tamper or a plate compactor so the surface ends up firm and even. Work around the pegs to leave those in place. If you have any sunken areas, you may need to add more base rock and compact. The site is now ready for your final surface.

LAY A GRAVEL PATH

There's nothing complicated about laying gravel; that said, don't cut corners on preparing the site and creating a good base. You will need edging to contain the gravel; there are many different options, but shown here are two of the more common—a wood edge and a brick edge.

YOU WILL NEED

Use your plan and the instructions on pages 245–247 to calculate quantities for

- base rock
- binding material (optional)
- gravel
- green pressure treated wood edging and fixing pegs OR bricks suitable for being used on the ground
- mortar (6:1 sharp sand to cement) for brick edging

Tools

- spray line
- line and pins
- tape measure and torpedo level
- hand saw
- club hammer
- tamper or plate compactor
- *cordless drill and drill bits
- *screwdriver and screws (1½in/ 4cm)
- rake
- spade and shovel
- brick trowel
- brick bolster or circular saw with a brick-cutting blade
- safety goggles

* not needed for brick edging

Over time, the weather (especially the wet winter months) and foot traffic will compact gravel. It's tempting initially to lay more than the recommended 1in (3cm) of gravel, but it's better to top it up every couple of years, as needed.

PREP YOUR SITE

First, mark out the area with spray line. To outline the area for your path clearly, fix pins at either end and strain a line tightly between them (see pp158–159). Use extra pins to navigate any bends, and make sure the line sits on the path's inside edge, then compact the ground (see pp160–161). I prefer not to use a weed-suppressing membrane beneath gravel paths; if you have a problem with pernicious perennial weeds, however, lay a membrane before adding the base rock layer.

EDGE YOUR PATH

There are many types of edgings to contain your gravel, in addition to the ones shown here. The height of your edging relative to the gravel depends on what sort of gravel you use (see p246): if using self-binding gravel, take it to almost the top of your edging; if it's pea gravel, I tend to keep it down to sit ¾in (2cm) below the edging to prevent it from spilling over.

WITH BRICK EDGING

Here, we are using a brick on edge finish, but you may choose to lay them flat to have a wider edge to your path or a cutting edge to a lawn.

1 After laying your base rock, set a line to the height you wish the top of your brick edging to sit. Mix 6:1 sharp sand and cement to make your mortar and use a brick trowel to place enough to sit your first brick beneath the string line. Make a small impression—what I call a valley—in the mortar mix. I like the mix to be pliable but not too sloppy. You may have to experiment when you first start laying bricks.

3 Continue laying bricks in the same way—butt jointing the bricks with mortar beneath each one but not between them. You could mortar between them, but if you have never laid bricks before, you will find this easier.

2 Position your first brick on the valley and tamp down with the handle of the trowel so that the top of the brick sits level with the string. It may take you a few goes to get the depth of the mortar correct.

4 If you are laying grass or have a bed up to the brick edging, cut away some of the haunch (the part of the mortar holding your brick at the sides) on the outside to allow for plant roots. Give the mortar a couple of days to set and then spread out the gravel (see steps 6 and 7 on p165).

> " *If you have a brick house, using a brick edge in your garden can be a lovely way of tying your space to the architecture of your house.* "

WITH WOOD EDGING

Wood boards are a more cost-conscious edging material than bricks. I use roofing lathes for pegs to hold the board since they are just the right dimensions.

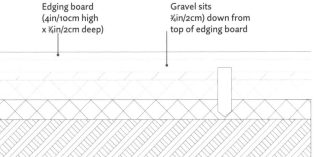

Edging board (4in/10cm high x ¾in/2cm deep)

Gravel sits ¾in/2cm) down from top of edging board

Supporting peg (10–12in/25–30cm long) knocked firmly into the ground

Base rock minimum 4in (10cm) deep

Compacted ground

FOR A GRAVEL PATH WITH EDGING BOARD

1 After prepping your site (see pp158–159), set your line for the height of your edging, making sure the line sits on the inside edge of the path. You don't have to create only straight paths with wood edging. If you want a curve, just use extra pins in your line to mark out any bends.

2 If you want your path to curve, you'll need to create flexible edging boards. Soak the wood the night before. Carefully saw one-third of the way through at intervals; if you need more of a bend, saw at closer intervals until it fits the curve. Be careful—they do snap occasionally.

3 Offer up the board to the height of your line or your set level. Secure by putting in the first peg on the outside edge of the path. Check that everything is vertical. Hammer in the first peg so that it sits 1in (3cm) below the edging board. Repeat, putting in a peg every yard (meter) along the length of the board. When putting in a curve, you will need to use pegs on the inside and outside to help guide the edging board into the curved shape.

6 Rake out a 1in (3cm) layer of gravel so that it sits about ¾in (2cm) below the top of the edging boards.

4 Fix the edging board to the pegs with screws, making sure you screw from the edging board out toward the pegs. Keep checking the levels as you go to make sure your edging board is still upright and sitting at the right height. Here, I am using the height on an existing bed, but you may need a line.

7 Backfill the soil to the outer edge of the boards. Once you've planted the bed, the board will "disappear."

5 To join two boards, put a section of wood across the join and screw in place. You may need to use a club hammer for support to push against. Remember to position your fixing board down from the top so you can hide it with soil or planting. With the boards fixed, spread base rock to a depth of 4in (10cm) using the edging as a guide. Then compact the surface with a plate compactor until firm and level (see p161).

> *Don't be tempted to lay your gravel too deep in one go. Let usage and rainfall settle it into place, then top up as necessary. You'll end up with a far better path.*

LAY SLAB PAVING

With a little guidance, it's not hard to build your own patio, terrace, or path out of paving slabs. As with any garden builds, be sure to spend the time on marking out and creating foundations. The steps shown here apply for all paving, from simple concrete slabs to materials such as granite, limestone, slate, porcelain, and sandstone.

YOU WILL NEED

Use your plan and the instructions on pages 245–247 to calculate quantities for

- base rock
- mortar mix for pavers (6:1 sharp sand to cement)
- mortar mix for pointing (6:1 soft sand to cement; 5:1 on lighter paving)
- pavers

Tools

- tape measure
- line and pins
- builder's square or 3-4-5 triangle
- spirit level and torpedo level
- stone saw (rented, if needed)
- bucket
- cement mixer (optional)
- rubber mallet
- spacers or tile spacers
- wire brush
- soft brush
- brick trowel
- pointing trowel

On your prepared base (see pp158–161), do a "dry run" area: set out some of your paving area temporarily to ensure you're getting the best possible layout and the best use of materials. You have your design on paper, but this is your chance to adjust to the actual slab sizes, which will give a better look overall and may reduce or remove the need to cut slabs. Always try to arrange slabs so that any cuts sit in discreet rather than in prominent places. If you need to cut slabs, rent a stone saw and wear a mask and goggles for protection.

When mixing mortar for pointing, it's a good idea to make a small batch first and let it dry out to check the final color. I tend to use white cement for lighter paving. Measure in buckets accurately for a consistent appearance; if not, colors may vary.

For a modern look, the smooth paving has a ¼in (0.5cm) pointing gap

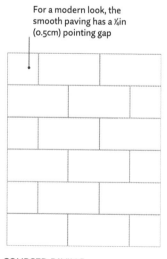

COURSED PAVING

For a traditional look, the textured random paving has a ⅜in (1cm) pointing gap

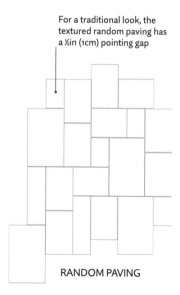

RANDOM PAVING

1 Once you're happy with the level of the base rock (see p158–161), check the level of your line—this is where the top of the pavers sit. Mix 6:1 sharp sand and cement. If you are laying a large area, it would make more sense to rent a cement mixer.

2 Lay a solid bed of mortar for the first paver. Initially, you may need to adjust the amount of mortar to lay—it can take a bit of getting used to. Once it's tamped down (with a rubber mallet or the handle of a hammer), the paver surface needs to sit at the line you've set.

3 Once you're happy with the size of the mortar bed, make sure the slab is sitting correctly with a torpedo level. As you lay your paving, keep using your line and level as a guide. The more slabs you lay, the less you will find this necessary.

4 Repeat for the remaining pavers, maintaining even spacing between the pavers by inserting spacers in the joints. Keep off the paved area for about a week to let it dry, then remove the spacers. Now you're ready to finish off with the pointing.

5 When you do the pointing, make sure it's a dry day. Here, I use a wet pointing mix for a clean, modern look. Alternatively, you could use a dry mortar mix, which you brush into the spaces. In both cases, you still need to work it into the spaces with a trowel.

6 If wet pointing, keep the paving surface clean of mortar by scraping it and wiping it down with a wet sponge, then protect the area with a tarp, if needed, while the mortar sets. If dry pointing, this stage is not necessary.

LAY BRICK PAVING

Bricks can be more versatile than large paving slabs, since their size makes them really useful for getting around curves in paths and terraces. They also offer lots of scope for various different patterns and optical effects and can potentially create an architectural link with your home.

YOU WILL NEED

Use your plan and the instructions on pages 245–247 to calculate quantities for

- ○ base rock
- ○ sharp sand (for bedding in bricks)
- ○ mortar mix (6:1 sharp sand to cement for fixing brick edging)
- ○ bricks (suitable for use on the ground)
- ○ kiln-dried sand

Tools

- ○ tape measure and spirit level
- ○ line and pins
- ○ wooden pegs
- ○ builder's square or 3-4-5 triangle
- ○ rubber mallet
- ○ shovel
- ○ wire brush
- ○ brick trowel and pointing trowel
- ○ long wooden plank
- ○ hand saw
- ○ metal rule and sharp-edged tool
- ○ brick cutter (rented if needed) or a brick bolster
- ○ sheet of hardboard
- ○ soft brush or broom
- ○ plate compactor

As with laying slab paving, there is a bit of maths involved in calculating materials and preparation work to be done to create your base (see pp158–161).

DO A DRY RUN

Once you've decided on a pattern, do a practice run by temporarily laying out some of the bricks in place. This will give you an idea of how it will all work and whether, by making any small adjustments to the size of your path, you can minimize the need for cuts. When you do start laying them, make sure you select bricks from across all the pallets you have, as this will give you a good variation of brick colors and a balanced look.

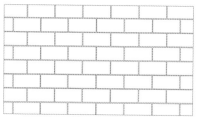

STRETCHER BOND

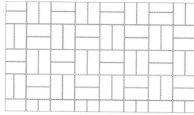

BASKET WEAVE

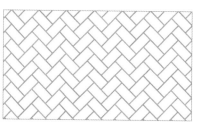

HERRINGBONE

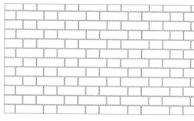

FLEMISH BOND

1 Prepare the site, marking out the area with line and pins, setting pegs to the level you want the finished surface to sit at, and compacting a layer of base rock in place to create a good base (see pp160–161). Recheck all the pegs within the site, adjusting as necessary to ensure that they are level and accurately reflect the height and slope you want for your finished paving.

3 As you lay your edging bricks, build a haunch (a ridge of mortar) on the outside edge of your area to hold the bricks in place. Trim away the haunch on the inside so that the infill bricks can sit snug. Let set for a couple of days.

4 Cut your wooden plank to create a screed board that is a bit wider than the space to be paved. Mark out and cut the L-shaped profile made by the brick edging so that the board sits on top of your brick edging. The section you cut out should be $\frac{1}{3}$–$\frac{1}{2}$in (8–10mm) smaller in depth than your bricks.

GO ON TO STEP 5

2 Lay a bed of mortar mix for your edging bricks (see p163). Adjust the depth of the mortar to ensure that the top of the bricks are all sitting level with the line you've set. Here, I'm using them flat, but you could use them on edge if you like. Here, I'm butt jointing as it's easier to start with, but you could leave a joint and point if you prefer.

" *If you have multiple pallets of bricks, use a mix of all of them in your paving to give you a balanced look and a good variety of brick colors.* "

5 Then, on top of the base rock, put down a bed of sharp sand about 1½in (4cm) thick. Make sure it looks a little higher than you need.

6 You want the sand to sit just above the base of the bricks—about ⅓–½in (8–10mm) up from the base of the bricks. Sit your prepared screed board over the brick edging and use it to "screed" (even out) the bed of sharp sand.

7 You will need to stay off the area after you have screed it, so work on a small section at a time. Work along the area and screed the sand so that it is the perfect depth to lay your bricks. You want the bricks within the area to sit about ⅓–½in (8–10mm) above the edging bricks.

8 You will need to cut some bricks before you start laying. Mark the cutting line with a sharp-edged tool, hold the brick firmly in place, and cut cleanly with a stone saw or a brick bolster (this needs practice to do neatly). Wear goggles and earmuffs. You will need to cut more at the end to fill any final gaps in your paved area.

9 Begin infilling the bricks. If you're using factory-made bricks, you can butt them up tight. Here, I'm using handmade bricks that have a little variation in size, so I'm leaving small pointing gaps. Plus I want to achieve a softer look. Check that the bricks are level and sit about ⅓–½in (8–10mm) proud of the edging. Do small sections at a time.

10 It's essential that you do the finishing off (steps 10–13) on a dry day. Make a small hole in the bottom of a bag of kiln-dried sand and pour the sand over to fill the joints between the bricks.

11 Use a soft brush to sweep the sand into place, filling the joints to the top. The sand will settle and may need topping up later.

12 Once all the bricks are in position, place a sheet of hardboard over the top and use a plate compactor to firm everything down. The brick paving should now sit level with the edging bricks.

13 I tend to leave a little sand on the top surface, so it all settles in with the weather.

" Laying brick paving is a great way to practice your landscaping skills. "

BUILD A LOW BRICK WALL

Once you've got the hang of laying bricks for edging paths or paving, you'll be able to build a brick wall. While you're still learning, don't build anything higher than 20in (50cm). Here, we are building what is called a "9-inch wall," which is two courses of bricks. This provides the strength needed to retain the soil behind.

YOU WILL NEED

Use your plan and the instructions on pages 245–247 to calculate quantities for

- base rock
- concrete (8:1 ballast to cement)
- facing bricks (suitable for facing walls) and any backing bricks (for course behind facing)
- mortar mix (4:1 soft sand to cement)
- weep holes and wall ties

Tools

- tape measure and spirit level
- line and pins or spray line
- builder's square or 3-4-5 triangle
- spade
- club hammer
- shovel, mixing tray, bucket, and wheelbarrow (or concrete mixer)
- block of wood for tamping
- hose
- brick trowel
- line blocks (to help make sure wall goes up straight)
- brick bolster or stone saw
- pointing trowel
- soft broom
- wire brush
- sponge and small bucket of water

For a wall under 20in (50cm) high, a footing of at least 12in (30cm) deep is usually fine, but a lot depends on the soil you're working with—you may well need to dig deeper to reach a hard, stable surface before laying the concrete footing. You also need to consider drainage as well as structural strength. Since this type of wall is visible from only one side, you can save money by using concrete blocks, or common bricks, for the rear side, and saving the good-looking facing bricks for the visible side.

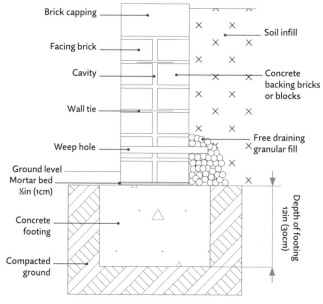

SECTION THROUGH THE BRICK WALL

3 Make sure your concrete level is flush with the top of your pegs, then use a block of wood to tamp the concrete footing down level. When the concrete footing is set (it'll take a few days), mark out the corners of the wall with a builder's square. Put pins in well past the ends of the wall and run a tight line between them where you want the wall to run; do the same for the line to mark the ends of the wall.

4 "Butter" one end of the brick. While you're learning, you might find it easier to hold the brick to the trowel as you lower it into place so the mortar stays put.

1 Use line and pins or spray line to mark out the base of your wall. Dig out your trench (roughly 12in/30cm) below ground level for the footing. Knock in wooden pegs to set the level for the base of your wall, making sure they'll sit just below ground level.

2 Mix the concrete and fill the trench, working along its length, up to the level of the top of the wooden pegs. I like the mix to be quite wet.

GO ON TO STEP 5 →

5 Set the line to the height of the first row of bricks, allowing for a mortar bed beneath it (usually bricks are 2½in [6.5cm] deep and the mortar about ⅓in [1cm] deep). Begin laying your bricks, pressing each buttered brick into the mortar on the footing and tapping it gently into place. Make sure it is straight and level.

7 Start by just laying a few bricks on each course, as the idea is to build up your ends (or corners) first. Aim to keep the mortar thickness even at ⅓in (1cm). After each row, use a spirit level to check that the vertical and horizontal levels are correct.

6 Lay bricks along the footing until you have the required length. Once this first row ("course") of bricks is completed, lay one brick at the end of the row, then begin the next layer by laying a corner brick across the two courses to create the stagger and the "stretcher bond." Repeat at the other end of the course.

8 Once you have built up the ends to their full height, begin filling in your facing and backing courses, using your line to check your levels. Set weep holes in your second course (see below).

9 To help drainage in the soil behind the wall, place a weep hole across both courses at the second course of the wall, above ground level, so water can drain through them. Place at intervals of 3ft (1m) along the length of the wall.

10 Wall ties literally tie both courses of bricks together, increasing the wall's stability (they're not strictly necessary for a wall as low as this). Put these in at regular intervals of 3–4ft (1–1.2m) along the wall.

11 If the joints start to drift slightly, don't worry too much unless it begins to look uncomfortable. If it does, just nip the end off a couple of bricks with the saw or bolster and you will soon have the joints realigned.

12 Keep surfaces clean of any excess mortar. If you don't finish the job in one day, cover your bricks with a tarp overnight if rain is expected.

FINISHING OFF

The top of the wall is exposed to the most extreme conditions, so it's vital to close it with a layer of bricks, stone, or paving slabs, which makes it look good, too.

1 Lay a few bricks on their edge at either end of your wall, checking that they are level. Use them as a guide to set a horizontal string line for the top of your wall. Continue laying bricks, working to your line. Tamp into position and keep checking levels.

2 If needed, use a bit of extra mortar and a small pointing trowel to ensure all joints are nicely filled and there are no gaps, if required.

3 Use a bricklayer's pointing trowel to smooth off and neaten the mortar joints between the bricks. Be sure that everything is left clean and tidy, then let set. It's a good idea to clean the wall down with a stiff or wire brush after 24 hours.

BUILD STEPS

Garden steps can be a lot more than a practical way of dealing with level changes. They can be used as clever design devices—for instance, by making an entrance appear much grander and more welcoming. Once you're confident about your bricklaying skills, building a few brick steps is a fairly straightforward process.

YOU WILL NEED

Use your plan and the instructions on pages 245–247 to calculate quantities for

- base rock
- concrete for foundation (8:1 ballast to cement)
- mortar mix (4:1 soft sand to cement)
- bricks (frostproof)
- kiln-dried sand

Tools

- line and pins or spray line
- hammer
- builder's square or 3-4-5 triangle
- carpenter's square and pencil
- spirit level
- tape measure
- wooden post, plank, and pegs
- spade
- shovel
- mixing tray or cement mixer (optional)
- brick trowel and pointing trowel
- stone saw (optional)
- tamper
- hand saw
- drill and bits
- screwdriver
- small soft brush

When putting in steps, always consider both their purpose and their aesthetics. Think about foot traffic—how often your steps will be used and by how many people—and the best materials for their location. For heavy usage, bricks or stone work well; for more occasional use, you could use sleepers.

PLANNING THE SITE

Measure the fall, or drop, of the slope where you want the steps to go (see pp20–21). Do a scale drawing to make sure the heights, depths, and widths of your materials will work. Remember to take into account mortar joints for bricks, which are usually around ½in (1cm).

Think about whether you'd like the brick treads to be wet or dry pointed (see p167). This will affect the final size of your treads. Personally, I prefer dry pointed infill as the treads weather nicely and moss can creep in at the edges, which gives them a timeless look. Wet pointing is a bit more time-consuming and often looks untidy if you haven't had a lot of practice doing it.

Start building only when you really understand how it will all work together with your chosen materials.

CALCULATING STEPS

You'll need to calculate how many steps will work comfortably for the incline (the vertical plane) and depth (the horizontal plane) available (see p244). Generally, I never create steps with treads smaller than 12in (30cm), and will always try to be more generous if possible. For the steps ("risers"), I try to work as close to 6in (15cm) as possible, although you could go up to 8⅔in (22cm) if necessary. Whatever size you go for, it's really important that all your risers are equal in size to prevent them from being a trip hazard. The width of the steps also plays an important part: narrow steps slow people down, whereas wider steps can become occasional seating. On average, I tend to make garden steps about 3ft (1m) wide.

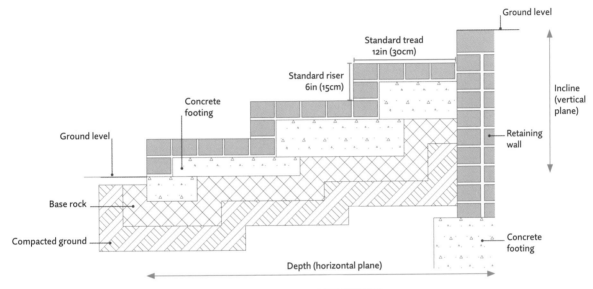

CROSS SECTION THROUGH THE BRICK STEPS

1 Lay out your bricks on site to double-check that your material works on site. This is the last chance to make any final adjustments. Use line and pins to peg out the foundations for the base of your steps.

2 You also need to know the heights at which your first step and last step will sit, so hammer in a wooden post at the middle of the steps close to the retaining wall. Use a carpenter's square and pencil to clearly mark on the heights of your steps (risers). This is a great way of keeping an eye on levels as you build.

GO ON TO STEP 3

3 Keep checking the measurements as you'll use these fixed reference points throughout the build.

5 Now set lines for the three sides of the U-shape for the height of your first brick course, double-checking that the front line is parallel to your back wall. Check that your corners are square by using the 3-4-5 triangle method (see p245).

4 Dig out the footing to around 12in (30cm) deep, knocking in wooden pegs to set the level (see p173, step 1). There's no need to dig the whole area, just a large U-shape where you want to lay the first set of bricks. Knock in pegs and create a concrete footing to that level (see p173). Let it set for a couple of days. This will form the base beneath your first step.

6 Start laying bricks in a U-shape on the concrete footing, with your retaining wall at the top of the U. Lay the bricks on a bed of mortar, buttering each end to hold them in place. Tamp down with the end of your trowel so the top of the bricks sits at the same height as your line and are level.

7 Lay a single row of bricks around the U-shape, following your string lines as a guide. Use a small spirit level to keep checking that your horizontals are level and your verticals are upright. Make sure corners are square.

9 Fill the area between the perimeter and the plank with concrete and use a shovel to make sure it's spread thoroughly. The concrete forms a firm base to lay the tread of your first step and the second riser.

8 Next, you'll need to make some shuttering—a temporary structure to contain the concrete base while it sets. Cut a plank of wood to fit inside the U-shape and position it about 4in (10cm) behind where you'll build your second riser. Make sure the top of the wood sits at the same height as your perimeter of bricks. Hammer in a couple of pegs and screw them onto the plank to hold it firm.

10 Use a short plank of wood that is wider than the step to tamp the concrete into place, getting rid of any air pockets and making it level.

GO ON TO STEP 11

11 When you've finished tamping, the concrete should sit neatly at the same level as your first line of bricks. Let the concrete set for a couple of days.

13 Now for your treads. Create a framework by putting down a layer of bricks at the front and side edges of all steps. Make sure this matches the level for your treads marked on your post, then saw off or knock in the post, as it's no longer needed.

12 Repeat steps 5–7 for your second step, setting your lines to the position of your second riser. Lay two layers of bricks in a U-shape behind your first step. Fill the base with hardcore, tamp it down, and then pour in concrete to form a second base that is level with the two layers of bricks, ready for your third step. There's no need for shuttering here as your retaining wall will hold in the concrete. Let it set. Repeat the process for your third and any subsequent steps. Keep checking the heights against those marked on your post and make sure everything is horizontally and vertically level. Take time and check, check, and check again.

14 Once you have laid all of your edging bricks, you can start on the infill. On top of the concrete on the lower step, lay a wet mortar bed of sharp sand and cement (4:1) to a depth of around ½in (1cm). Place a layer of bricks for the treads in the mortar bed. As you're working, tamp down each brick to make sure it's level and at the correct height. Repeat for the treads of the other steps.

15 Once the bricks have set, pour kiln-dried sand over the top of the treads to fill the gaps between the bricks. Make sure that you do this on a dry day.

17 Keep off the steps for around a week so that everything sets firmly.

16 Then use a soft brush to sweep it into place, filling joints to the top. The sand will settle and may eventually need topping up.

CONCRETING POSTS

Knowing how to fix wooden posts into the ground with concrete is a really useful skill that you can use on lots of garden building projects. It's also easy to do, so you'll soon be able to put your newfound skills to use on putting up fencing, building an arbor, or creating archways.

YOU WILL NEED

Use your plan and the instructions on pages 245–247 to calculate quantities for

- wooden posts 4 x 4in (10 x 10cm)
- concrete materials (8:1 ballast to cement) or ready-to-use mix

Tools

- tape measure
- line and pins
- carpenter's square
- pencil
- spade
- post-hole digger (optional, but really useful)
- crowbar (optional)
- hand saw
- spirit level
- concrete mixer (optional)
- mixing tray
- shovel
- wood battens for tamping concrete and propping posts
- screwdriver and screws (for temporary fixings)

The choice of wood will affect the overall look of your garden but also its longevity. Softwood (which for outside use should always be bought as treated wood) is cheaper than hardwood, but it won't last as long. Green oak is also relatively cheap but will move, twisting and cracking as the weather changes, whereas dried oak will not move as much. Be mindful that when you use wood outside, it will react to the environment and change over time. Usually it will turn silver-gray unless you treat it each year. You can also stain or color the wood to suit your scheme. Planed wood is smooth and modern looking; sawn is more rustic.

1 Mark the position of your posts. Make the holes roughly 12 x 12in (30 x 30cm) wide to allow space for the post plus the concrete around it. Dig holes to a depth of 24in (60cm) using a spade. A post-hole digger and a crowbar can also come in handy.

4 As you're filling the hole, use a wooden batten to tamp in the concrete and dispel air pockets. Once the concrete reaches 3–4in (7.5–10cm) below ground level, give it one last tamp.

2 Next, mix your concrete (or use a ready-to-use mix). I tend to make it on the wet side, so that it is easier to work in around the posts and reduces the potential for air pockets. Offer the post into the hole and bring it to vertical, using a spirit level to check.

3 Hold it in place (or get someone to help) and start to fill the hole with the concrete mix, taking care not to move the post out of position. If you're putting in more than one post, make sure the tops are all level or step down on a slope (see p243).

5 Support the post in its position using four wooden battens as props while the concrete sets (within 24 hours). Put screws part way into each side of the post, just above the battens, as a temporary fixing. When the concrete is set, remove the screws and battens. The post is finished.

ERECT FENCING

Putting up fencing isn't difficult, and it's possible to do it by yourself—although if you can get someone to help, it'll be even easier. Once you've mastered installing posts, fixing panels, and keeping everything upright and in a straight line, you'll be able to put up trellis, install a decorative panel, or build a compost bin.

YOU WILL NEED

Use your plan and the instructions on pages 245–247 to calculate quantities for

- ○ fence posts and panels
- ○ finials (optional)
- ○ concrete materials (10:1 ballast to cement) or ready-to-use mix

Tools

- ○ spray line
- ○ line and pins
- ○ builder's square or 3-4-5 triangle
- ○ tape measure
- ○ spade
- ○ shovel
- ○ club hammer
- ○ mixing tray
- ○ concrete mixer (optional)
- ○ drill and drill bits
- ○ screwdriver and screws (for panels and temporary fixings)
- ○ wood battens for propping posts
- ○ blocks of wood for wedging
- ○ spirit level
- ○ hand saw
- ○ screw eyes, 0.06in (1.6mm) galvanized wire, wire cutters, and pliers (optional)

Fencing can be a really important part of your design, especially in a small garden area, so be sure to choose materials carefully. You can either make your fencing a feature of the garden or use it as a support for climbing plants so it disappears into the background.

WHERE TO PUT THE FENCE?

Check exactly where your boundary line falls and whether the local planning authority has any height restrictions. If your fence is on a slope, make sure the rise or fall of the panels is evenly spread or is stepped in an attractive way (see p243).

> " *For me, the gardens that have real atmosphere have either really cool boundaries that are integral to the design or fences that completely disappear because they're covered in plants.* "

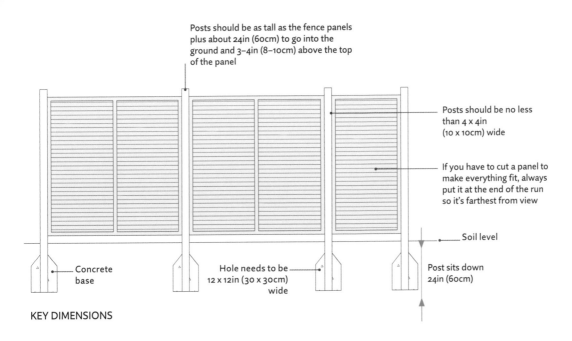

Posts should be as tall as the fence panels plus about 24in (60cm) to go into the ground and 3–4in (8–10cm) above the top of the panel

Posts should be no less than 4 x 4in (10 x 10cm) wide

If you have to cut a panel to make everything fit, always put it at the end of the run so it's farthest from view

Soil level

Concrete base

Hole needs to be 12 x 12in (30 x 30cm) wide

Post sits down 24in (60cm)

KEY DIMENSIONS

1 Clear the fence line of any trees, shrubs, plants, or debris. Use spray line first then line and pins to mark the fence line. Be sure to keep the line tight.

2 Dig out the hole for the first post. It should be about 24in (60cm) deep and 12 x 12in (30 x 30cm) wide. Follow the instructions for concreting posts (see pp182–183).

GO ON TO STEP 3

4 Offer up a fence panel to your finished height. Hold in place with blocks of timber as wedges. Using a spirit level, double-check that everything is level and upright. (You can do this job by yourself, but if you have not done it before, it may be easier to get someone to help you.)

3 Never dig more than one post hole at a time. That way, you know all the panels will fit neatly and measurements won't get out of kilter. Once you've concreted the first post in place, partly screw battens in place to support the post while the concrete sets.

5 To fix the panel to the first post, predrill holes into the upright of the panel, top and bottom. Then screw the panel in place. I tend not to screw all the way into start with, just in case I need to make any adjustments.

7 Fix the panel to the post with screws. Check that the post is upright and then backfill around the post with concrete, as before, and support with wooden battens while the concrete sets.

6 Dig the next hole and put the second post into position. Make sure the posts are level (unless you are on a slope and need to step the posts up or down; see p243). The top of the posts should sit 3–4in (8–10cm) above the fence panel.

8 Continue in this way, until all the posts and panels are in position. Keep the props in place until the concrete is set. Tighten the screws to hold the panels firmly to the posts. If you want to grow climbers up the fence once it's finished, now would be the best time to put in supports. Attach screw eyes to the fence posts and strain wires tightly between them.

BUILD A PLANT SUPPORT WITH ROPE SWAGS

Once you have mastered the art of concreting posts, creating a framework for supporting climbing plants, such as roses, clematis, and honeysuckle, is a straightforward building project.

YOU WILL NEED

Use your plan and the instructions on pages 245–247 to calculate quantities for

- pressure treated wood posts 4 x 4in (10 x 10cm) wide x 9ft (2.7m) high
- concrete (6:1 ballast to cement) or ready-to-use mix
- 1in (3cm) diameter manila rope (allow 10¾ft [3.3m] rope between posts 10ft [3m] apart)

Tools

- pencil and tape measure
- drill and flat, wood drill bit 1¼in (3cm)
- line and or pins spray line
- spade or post-hole digger
- shovel
- mixing tray
- concrete mixer (optional)
- paint or wood stain (optional)
- ladder
- spirit level
- wood battens for props
- gaffer tape
- string
- screwdriver and 5in (12.5cm) screws
- screw eyes, 0.06in (1.6mm) galvanized wire, wire cutters, and pliers (optional)

You could, of course, go for a wood archway, which could work wonderfully along a path, but be mindful, as I find a common mistake people tend to make is placing their archway too close to paths and not allowing enough height for the plants to inhabit the archway, which ultimately can make them uncomfortable to move through.

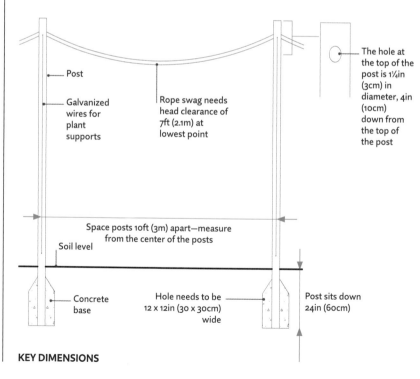

Post

Galvanized wires for plant supports

Rope swag needs head clearance of 7ft (2.1m) at lowest point

The hole at the top of the post is 1¼in (3cm) in diameter, 4in (10cm) down from the top of the post

Space posts 10ft (3m) apart—measure from the center of the posts

Soil level

Concrete base

Hole needs to be 12 x 12in (30 x 30cm) wide

Post sits down 24in (60cm)

KEY DIMENSIONS

3 Repeat for your remaining posts. Make sure the drilled holes at the top of the posts all face the same way so that rope can be threaded through. Let the concrete set overnight.

1 First, mark out the position of your hole. Measure 4in (10cm) down from the top of the post. Drill the hole, being careful not to damage the other side of the post. Make sure the hole is centered and in the same position for all posts. Repeat for the second and any subsequent posts.

2 Mark on the ground the position of each post using spray line or line and pins. Follow the instructions on digging down and concreting posts (see pp182–183) for your posts. Remember to partly screw in battens to support the posts while the concrete sets. These can be removed after 24 hours.

4 Pressure treated wood is designed for outdoor use, but you might want to paint or stain the posts. If so, do that next. It's good to repaint them every couple of years.

GO ON TO STEP 5 →

5 Simply use gaffer tape to stop the ends of the rope from fraying. Then to hide the tape, bind over it with string.

6 Using a ladder so you can reach, thread the rope through the holes. You will need to stand back and adjust the swags until you're happy with the look. Secure in place, working at an angle to put a couple of screws through the rope and into the post to hold it.

7 Repeat this process for the third and any subsequent posts. Cut off any excess rope, if necessary, and bind the end with gaffer tape and string. For a double swag, repeat the same process as before, checking that you're happy with the way the rope hangs. Lay the second rope on top of the post, ensuring it stays above the lower rope.

> *Creating a framework for supporting climbing plants, such as roses, clematis, and honeysuckle, is an effective but straightforward building project.*

10 Use garden twine to secure climbers to the wires. You'll need to add more ties as the plants grow. Remember to check that they remain secure, particularly during windy weather when plants may come loose.

8 Screw the second rope onto the top of the post to hold it in place. Make sure you screw through the center of the rope to the center of the post. Also make sure the end of the rope sits at the far edge of the post.

9 I find it easiest to train plants up the posts using vertical wires. Attach a screw eye near the top and bottom of the posts on each face, cut wire to length, and attach between the two, pulling and winding it taut.

11 Once your climbers have become established on the wires, you can train them between the posts by tying them to the rope. The horizontal growth will encourage more flower.

BUILD A RAISED BED

Here, I show you how to build a basic raised bed, but you can tailor the dimensions and style of yours so that it suits your own requirements. You may want your bed to sit higher off the ground or use thinner boards if you are limited for space. Just adapt the principles for your needs.

YOU WILL NEED

Use your plan and the instructions on pages 245–247 to calculate quantities for

- pressure treated wood posts 4 x 4in (10 x 10cm) wide cut into 41in (105cm) lengths
- pressure treated sleepers, cut to 47¼in (120cm) and 72in (183cm) long (they are 8in [20cm] deep x 2in [5cm] wide)
- concrete (8:1 ballast to cement) or ready-to-use mix

Tools

- tape measure
- line and pins or spray line
- builder's square or 3-4-5 triangle
- spirit level
- pencil and carpenter's square
- hand saw
- spade
- shovel and mixing tray, or concrete mixer (optional)
- club hammer
- drill and 20mm flat, wood drill bit and 8mm drill bits (for boltholes)
- 6in (15cm) coach bolts and 4in (10cm) screws
- socket head for drill or socket set (to tighten the coach bolts)
- rope, bamboo canes, or wood battens for mesh support (optional)

Prepare your raised bed site: it needs to be clear of perennial weeds and have free-draining soil and an even base. Your style of bed is up to you; here I have used taller corner posts to support netting, but if you don't want to do this, you could fix 2 x 2in (5 x 5cm) pressure treated posts, concreted into the same position but fixed below the sleeper sides. If you'd like to paint or stain the wood of your raised bed, do this before constructing the frame.

DECIDE ON THE SIZE

Always make sure the width of the bed is narrow enough for you to reach across to the middle easily from both sides for planting and weeding. Here, the bed is about 47¼in (120cm) wide. The longer side of this raised bed measures 60in (150cm), but if you go any bigger than 94½in (240cm), you may need to put in extra supports along its length.

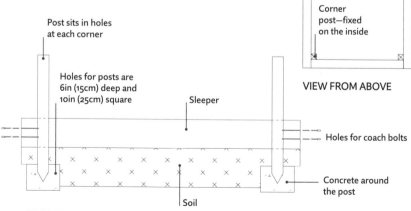

Post sits in holes at each corner

Holes for posts are 6in (15cm) deep and 10in (25cm) square

Sleeper

Corner post—fixed on the inside

VIEW FROM ABOVE

Holes for coach bolts

Concrete around the post

Soil

SIDE VIEW

1 Before you start to build, measure and mark the sleepers to the required length, then cut them to size. Next, using line and pins, set out the area and levels for your new bed.

2 Offer up your end sleeper to your line and hold it in place with soil. Dig out the site for the corner post to roughly 6in (15cm) deep and 10in (25cm) square. Next, lay out the other three sleepers in turn, holding them in place with soil and digging a hole for each corner post.

3 Use a spirit level to make sure the sleepers are level and their tops sit flush with your line. At each corner, drill two holes for coach bolts (one near the top and one near the bottom) through the sides of the long sleepers to fix them to the ends of the short sleepers. Next, bolt the corners together with coach bolts.

4 Drill a ¾in (2cm) hole into the top of each corner post (see p189). Lower the corner posts into their holes. Check that the tops of all four corner posts are at the same height (see p243), and then secure in place with 4in (10cm) screws.

5 Check the levels again, then concrete in the corner post holes to hold the posts in place. Once the concrete is set, everything will be held firmly in place. (If you're building taller beds, add your additional layers of planks at this stage.)

6 Fill the bed with a mix of garden soil and compost, level off, and tamp down ready for planting. Insert rope, cane, or wood battens into the holes on the posts and secure in place if necessary. These crossbars can support nets or fleece to protect crops.

BUILD A WATER FEATURE

Creating a focal or destination point using water in your garden provides another layer of interest and mood; it's also a fantastic way to attract wildlife into your garden. This basic trough filled with water-loving plants is a great "starter" project, as it requires no plumbing or digging.

YOU WILL NEED

- watertight container, such as a galvanized water trough
- spirit level
- air-pot cut to size or aquatic plant baskets
- cable ties (optional)
- bricks (optional)
- aquatic soil
- bucket or large trug
- water plants, including floating, oxygenating, and marginal plants
- small pebbles or gravel
- cobblestones (optional)

I'm using a reclaimed galvanized trough, but you can use all sorts of containers for this project, as long as they are watertight. Try your local garden center, or for a container with a bit more character, you could visit a local reclamation yard or search eBay for something vintage.

AQUATIC PLANTS

When buying plants, make sure they are suitable for the space you have and choose a variety of shapes and textures that work well together. Oxygenating plants are a must to keep the water fresh and its ecosystem in balance. Ultimately, the balance of plants to water will help keep your water clear. It may take a little time and adjustment to find the right balance. You can use bricks to raise the plants in pots to the right level for them to thrive.

> *Make sure you are happy with the position of the trough before planting it up. Once it's full of water, it will be a real pain to move.*

1 Select a level, firm piece of ground as a site for your container. Take your time to look at it from all angles and do try it out in various places. Once you're sure you've found the best site, use a spirit level across the top of the container to make sure it is level in both directions.

2 I often buy trees in air-pots (made from perforated plastic) and then recycle these into containers for water plants. Simply cut the perforated plastic to size and hold the pot together with cable ties. As an alternative, you can buy aquatic planting baskets.

5 Cover the surface of the soil with a ¼in (0.5cm) layer of gravel or pebbles to hold the soil in place.

3 Arrange your pots or baskets in the bottom of the water trough. Bear in mind the eventual height and size of each of the plants to figure out how best to position them. You can raise some of them up by putting bricks beneath them if needed.

6 Water the plants and then gradually fill the trough, letting it overflow slightly to wash away any stray soil floating on the surface. If you have an existing water feature or a rain barrel, add a bucketful of that water to help establish healthy microorganisms in your new water feature.

4 Use aquatic soil to half fill the containers, then add your plants and top up with soil. Keep the soil about 1½in (4cm) down from the top of the container.

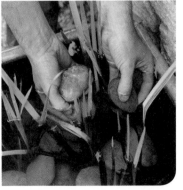

7 You can also put a layer of cobblestones on top. They come in a mixture of colors and sizes, and their smooth texture can really enhance your water feature.

BUILD

PLANT YOUR GARDEN

Nothing beats the excitement of finally getting your hands on your new plants.

INTRODUCTION

The hard landscaping features are in place; now, it's time to fill your garden with all those beautiful plants you've chosen. Planting is the most exciting part of the garden-making process for me—it's an opportunity to bring the space to life with form, color, and texture.

I love every stage of creating a garden, but putting in the plants has got to be the most fun part. After all the practicalities and building work, when I finally get my hands on the green stuff, it always puts a smile on my face. Even when I'm doing show gardens, I can't wait for the plants to turn up.

Many people get quite anxious when it comes to planting, but in reality, the only bit of soft landscaping you need to get close to perfect is your trees and shrubs, as hopefully they will be there for the long term. As for the rest, it doesn't matter if you're shuffling stuff around for the next 20-odd years.

Inevitably, you'll get some things right and you'll get some things wrong, so don't worry too much. As long as you make sure to put your plants in the right location, with healthy, nutritious soil and enough room to grow, everything else should come with time. So just try to learn from it all and enjoy the process.

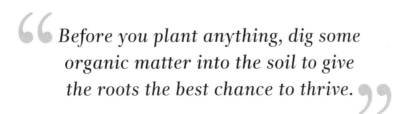

Before you plant anything, dig some organic matter into the soil to give the roots the best chance to thrive.

BASIC TOOLS FOR SOFT LANDSCAPING

You'll need a good selection of cutting and digging tools to keep your garden looking its best. I tend to check all my tools over during the winter months, then give them all a good clean and a wipe with an oily rag.

6. Knife

8. Hand trowel

1. Loppers

2. Hoe

10. Hand fork

3. Lawn edging tool

4. Garden shears

11. Pruning shears

5. Pruning saw

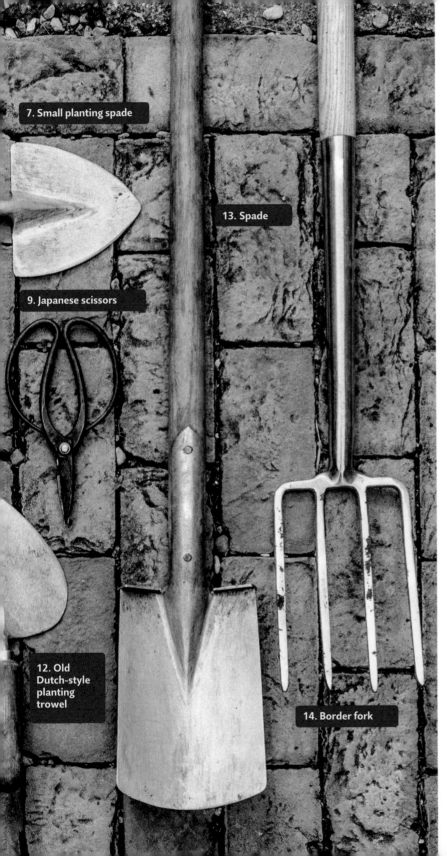

7. Small planting spade

9. Japanese scissors

13. Spade

12. Old Dutch-style planting trowel

14. Border fork

CUTTING TOOLS

1. Loppers for pruning **2. Hoe** for lifting weeds and working the surface of the soil **3. Lawn edging tool** for keeping lawn edges neat **4. Garden shears** for cutting hedging or long grass around trees **5. Pruning saw** for pruning thick branches **6. Knife** for cutting twine and taking cuttings **9. Japanese scissors** for cutting flowers, deadheading, and light pruning **11. Pruning shears** for pruning and cutting back

DIGGING TOOLS

7. Small planting spade for 2-3qt pot-sized plants **8. Hand trowel** for small planting **10. Hand fork** for loosening soil when weeding **12. Old Dutch-style planting trowel** a heavy-weight trowel for planting and loosening soil **13. Spade** for digging, with treads at the top to protect boots **14. Border fork** for lightly digging, lifting, turning, and aerating soil

OTHER (not shown)

Rake for creating the right tilth of the soil and for treading and tamping when laying sod

PREPARING THE GROUND FOR GRASS

Grass is relatively cheap to install and maintain and it's a great way of "carpeting" larger areas of your garden area. Whether you want to sow grass seed or lay sod (see pp204–205), the preparation is pretty much the same—you need an even base and a fine tilth of soil.

YOU WILL NEED

- line and pins or spray line
- tape measure
- builder's square or 3-4-5 triangle
- spade
- tiller (rented, if necessary)
- fork
- large rake
- roller (optional)
- spirit level

> *If you've got poor soil, you're probably best off buying some good-quality topsoil to give your new lawn a head start.*

If you want a lawn in your garden area, the first thing to ask yourself is what you want to use it for and how much time you want to spend on maintaining it. There are various mixes of grass seed and different sorts of sods, each one designed with a particular purpose. If you do a little research, you'll soon find one to suit your needs, be it a low-maintenance or a fine-textured sod.

For me, the key thing with any lawn is to have at least 4in (10cm) of decent topsoil. The ground beneath needs to be free draining. The surface should be firm and even and ideally have a very slight gradient for rain to drain easily.

WHICH TO CHOOSE—SOD OR SEED?

The big differences between laying sod or sowing grass seed are time and money. If you lay sod, you get an instant lawn, whereas if you sow from seed, you have to wait for it to grow and deal with weeds; however, seed is considerably cheaper.

WHEN TO LAY SOD OR SOW SEED?

You can lay sod year-round in good weather conditions, but for me the best time is spring or late summer. The best time to seed a lawn is in early fall or late spring.

GETTING THE GROUND READY

Choose a fine day to do the heavy work of prepping the soil ready to receive the grass seed or sod.

1 Mark out the areas you want to sod. Roughly dig over the area to break up the soil. If it is a small area, you can use a fork, but for larger areas you will need to use a tiller. Make sure it is clear of weeds, especially perennial weeds, as well as large stones and any debris.

2 Knock the soil through with a fork to roughly level the area, removing stones, roots, and any big clods of earth.

3 Use the widest rake you can get hold of to level the surface of the soil, taking out any dips and bumps.

4 Lightly tread all over the site to make it as flat and even as possible; you could use a roller if you prefer. Rerake the area one last time to create your final level. You are looking to produce a fine tilth (something that looks a little like bread crumbs).

HOW TO SEED A LAWN

The best time to seed a lawn is in early fall (ideally September when there's warmth in the soil and enough time for a decent germination) or in late spring. If you sow over winter, the chances are you'll get poor germination; if you sow in summer, the young seedlings can get too stressed in the heat and can easily dry out. People often oversow, but to get a healthy lawn, try to avoid doing this.

1 The grass seed packet should indicate how many grams per square yard (meter) you should sow (see pp245–247). It helps to use line and pins to mark out 1 sq yard (meter), then measure out the seed and throw handfuls of it (known as broadcasting) as evenly as possible over the area. Once you get the hang of roughly how much you need per sq yard (meter), you should be able to do it by eye. Alternatively, you can buy or rent a grass-seed spreader to help you.

2 Lightly rake or roller the area to help key the seed into the soil, then stay off the grass until germination has taken place—anything from 5 to 30 days.

LAY SOD

Whether you dream of bowling-green-like clipped grass or more meadowlike areas of longer grass with pathways mowed through, laying sod will realize your dreams quicker than sowing seed. Anyone can do this— it's a straightforward and really satisfying job, and you'll be rewarded with an instant lawn.

YOU WILL NEED

Use your plan and the instructions on pp245–247 to calculate quantities for

○ sod

Tools:

○ large rake
○ sharp knife
○ scaffold boards or wooden planks
○ roller (optional)
○ lawn edging tool
○ watering can or hose

Your sod will come rolled up in sections, but make sure your soil is ready before you have it delivered. If you'll be laying them quickly, they'll be fine left stacked; if not, find an area to lay them out and water them if needed. Think of the sod as carpet tiles of sorts, and always work with as big a piece as possible. When I'm laying sod, I always lay it in lines across my view. Though the edges of the individual sod pieces will soon meld together, laying them across your line of sight improves your view from day one.

Always wait until the grass is established before cutting it for the first time. Even then, I never cut lawns lower than 2in (5cm). Cutting lawn too short is a common problem as it can put the grass under stress.

1 Once the ground is ready (see pp202-203), begin by laying the edges of the lawn, creating a sort of "framework" around the shape you want your lawn to be. Use scaffold boards to work off so you avoid treading on the soil you have prepared. Make sure the sod pieces are level and even. Tamp them into place using the back of a rake.

2 After you've laid the edges, fill in the area, working across the view from your house, not along it. Use staggered joins (in a bricklike pattern), so once you have laid your first row, cut a sod piece in half with a sharp knife to start the second row.

4 As you progress, by moving and treading on the scaffold boards along, you'll find the sod pieces are tamped down. When you've finished laying sod, go back over the site, again using the scaffold boards, to make sure everything is tamped down neatly. Finally, give it all a really good soak.

Depending on the time of year, you'll need to water your lawn a few times as it gets established. Keep off the grass, as feet can damage young growth and make the lawn uneven.

3 The key to laying sod well is the way you pull and butt the pieces into each other. If it's not tight enough, when the grass dries and contracts—which it can do—you'll end up with gaps. So push each piece against its neighbor so it sits really nice and tight.

PLANT TREES AND SHRUBS

Whether you buy trees and large shrubs for the garden as root balls or as bare-root stock, it's important to plant them in a hole with the right shape and at the right depth. Most shrubs won't need support, but once they're in the ground, trees will need support via the use of cross or vertical staking, taking care not to damage their roots.

YOU WILL NEED

- spade
- pruning shears
- scissors
- large bucket of water or hose (depending on the size of the plant)
- mycorrhizal fungi (optional, to encourage good root development)
- spirit level
- peat-free compost, grit, or well-rotted manure (optional)
- claw hammer
- nails
- wooden stake (diameter of 3in/7.5cm), pointed at one end
- club hammer
- rubber ties
- hand saw

> " *The key to good tree planting is to prepare the hole correctly, and to not plant your tree too deep.* "

Trees and shrubs are sold with root balls (in containers or wrapped in fabric) or with bare roots. Those in containers can be planted at any time of year, whereas cheaper bare-root trees can be planted only from winter to early spring, while they're dormant.

Always give trees and shrubs enough space to grow to their eventual height and spread. Dig a decent-sized, square-shaped hole, which encourages the roots to grow outwards, and don't plant too deep—keep the top of the root ball just above ground level.

1 Mark the location as per your scale plan if you're using one. If you are planting a tree or shrub with a root ball, measure the root ball and dig a square hole big enough for the roots plus an extra 6in (15cm) on all sides.

2 Check the size of the root ball against your spade to ensure that the hole is the right depth. For bare-root stock, clip off any straggly or damaged roots and make sure the hole is big enough for the tree or shrub to sit in comfortably.

3 If planting a root ball tree or shrub, soak the plant's roots with water, then remove the plant from its container or fabric wrapping. There's no need to presoak a bare-rooted tree. Sprinkle some mycorrhizal fungi onto the roots over the hole, especially if your garden is on a new site.

4 Place the tree or shrub into the hole so that the base of the stem sits just above the ground; it will settle a little so resist planting it too deeply. Stand back and make sure the trunk is vertical.

5 Gradually fill in the soil around the roots, firming the soil as you go. If the soil taken out of the hole is thick and cloddy, I sometimes mix in compost and grit (or even some well-rotted manure if the soil is really poor), before putting it back into the hole. This helps keep the soil finer and reduce air pockets within it.

6 Your new tree or shrub will need support. For root ball plants, use a cross stake (at a 45° angle). Hammer the stake into the ground, then secure it to the tree with a rubber tie; tight, but not too tight. Saw off the excess stake with care. Water the newly planted tree or shrub routinely during its first year or two, especially during spring and summer, to give it the best chance of success.

VERTICAL STAKING

A bare-root tree also needs support, but the stake can be parallel with the stem (not at 45° as shown above). That said, be sure not to damage the roots when you hammer in the stake. Once the stake's in place, you can backfill the hole.

PLANT PERENNIALS

Perennials' quick and vigorous growth mean they can be hungry plants and can deplete the soil easily. In order to achieve the best results, it's important to prepare the soil well, provide the right conditions, and be mindful of the weather, ensuring that the plants don't dry out while they are establishing themselves.

YOU WILL NEED

- peat-free compost or well-rotted manure
- grit
- border fork
- spade
- rake
- hand trowel
- mycorrhizal fungi (optional)
- bucket of water

> *This is the best bit for me—I love setting out my plants and making final tweaks to my planting plan.*

All new plants, including perennials, need to be able to easily grow their roots outward to take up water and nutrients. To give them the best start, make sure the soil is in "good heart," by which I mean it's healthy and fertile. It's a good idea to dig in some organic matter, such as compost or well-rotted manure, about a month or two before you start planting. If drainage is an issue, you might want to put in some grit, too.

Container-grown perennials can be put in at any time of year—apart from icy or very hot, dry conditions—but if you plant them out in spring or fall, they are less likely to get stressed.

1 With your planting beds marked out, you can prepare your soil. Start by digging over the beds with a fork or spade, clearing any weeds, old roots, and large stones or clods as you go. Then work in some organic matter so that it's even and feels good to plant into.

2 Arrange the plants in the bed still in their pots, referring to your planting plan as needed (see pp146–147). Double-check their labels to find out how big they will grow in terms of height and spread, and space them out accordingly. This is the time to make any final adjustments, so make sure everything feels right.

4 Remove each plant from its pot and tease out the roots to loosen them.

5 I tend to dunk each plant's roots into a bucket of water until air bubbles stop appearing. Then, it's ready to plant.

3 Dig holes for each of your plants. Take your time when preparing them. Be generous with the size; make them slightly bigger than the root ball so that when you put the plants in, you don't have to ram them in place. If your garden is on a new site with poor soil, add mycorrhizal fungi to the holes at this stage.

6 Position each plant in its hole and backfill with soil, using your hands to firm in to avoid any large air pockets. Water in the new plants well. Keep an eye on the weather conditions and don't let the plants dry out while they're establishing themselves.

PLANT BULBS

The arrival of spring color is a welcome sight. For me, not only do bulbs provide another layer of interest in the garden, but their arrival really kicks off the year. Remember, though, that bulbs don't just add spring color—there are plenty of bulbs that provide summer interest, too. I love them popping up through and around other plants.

YOU WILL NEED

- lawn edging tool
- sturdy spade or trowel
- long-handled bulb planter (optional)
- watering can
- pots
- grit
- peat-free compost

" Plant colorful container displays of tulips near the house, and you're more likely to enjoy them every day in spring. "

BULB PLANTING DEPTHS

No matter which planting method you use (see opposite), bulbs need to be planted at different depths depending on their type. I've shown a few examples below, but as a general rule, a bulb needs to be planted at two to three times its depth. If you are unsure, the packet you buy them in should specify planting depths you need to follow.

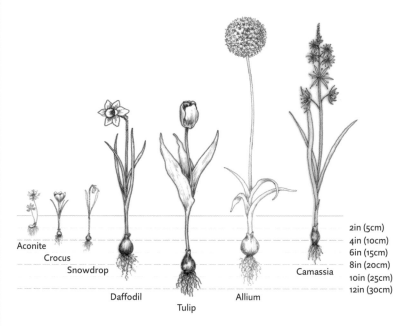

Aconite
Crocus
Snowdrop
Daffodil
Tulip
Allium
Camassia

2in (5cm)
4in (10cm)
6in (15cm)
8in (20cm)
10in (25cm)
12in (30cm)

DEPTH OF PLANTING FOR BULBS

PLANTING BULBS AMONG GRASS

If I'm planting bulbs to naturalize and spread themselves as they would in the wild—such as *T. sylvestris*, my favorite native tulip—I tend to plant a few larger groups and then scatter bulbs out from that group.

1 Figure out roughly where you'd like to plant your groups of bulbs and the look you want. Stand back from the site and imagine looking at them. Then use a lawn edging tool to cut out the area(s) of grass.

2 Lift the grass and carefully set it to the side so you have clean soil below. Dig down to the depth required for the bulb you are planting (see opposite).

3 Position the bulbs at the required depth and spacing—here, I've used *Narcissus pseudonarcissus*. Then carefully replace the grass, with its soil, and gently tamp down. Water well.

PLANTING INTO POTS

Container planting gives you flexibility—I tend to position pots where they can be viewed from the house or where I will pass them on the way to the car.

1 Choose a clean container that is a suitable size for the number of bulbs you are planting. If you are unsure about spacing, check the packet instructions. Mind you, I do tend to plant a little closer in containers.

2 I plant most of my bulbs in a free-draining mix of one part grit to three parts peat-free compost. Fill the pot to the level at which the bulbs need to sit.

3 Plant the bulbs at the correct depth (see opposite), then cover with more compost mix and a handful or two of grit.

PLANT "IN THE GREEN"

I love to use this method for my borders, planting bulbs up in pots in the fall, then planting them out into the ground when the stems are showing (usually early spring). That way I can really understand how they will look.

ENJOY

> " *For me, there is something to be enjoyed in your garden every day of the year.* "

INTRODUCTION

I love watching the seasons unfold in the garden, from the fresh greens of spring to the bareness of winter. A garden is the best place to unwind and reconnect with nature. It's pretty special having a garden. As you learn to care for it, you start to enjoy what each season has to offer.

A lot of new gardeners are a little bit fearful of what they need to do, the timings of everything, and how much work it involves, but it's really not that difficult. If you spend a few minutes each day puttering about, you'll start to see that there's a rhythm to the gardening year, and there are periods when the garden will need more or less of your time. Learn when you need to put a bit of extra work into your garden, and you'll make the rest of the year a lot easier.

In this chapter, I've listed how I go about caring for my garden, but in reality, the month you do a particular task doesn't have to be set in stone. Some tasks are mentioned a few times across different months, but they may only need to be done once. Much can depend on where you live, too; spring, for example, starts earlier in the south than in the north. So just use these lists as a framework for garden tasks and don't get hung up on exact timings.

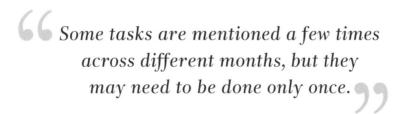

Some tasks are mentioned a few times across different months, but they may need to be done only once.

JANUARY

January can often feel like the worst month in a garden. The cold weather and wet ground might mean that you're better off staying indoors. However, there are still a few jobs that can be done, such as preparing empty beds and putting in bare-root plants. You can also take advantage of time indoors by planning the gardening year ahead.

QUICK CHECKLIST

- ○ **Clear leaves and debris.**
- ○ **Dig over empty beds.**
- ○ **Plant bare-root trees, shrubs, and fruit bushes.**
- ○ **Protect plants, clearing snow from them if necessary.**
- ○ **Feed birds and wildlife.**
- ○ **Order seeds.**

Cyclamen

IN YOUR BORDERS AND VEGETABLE GARDEN

- Keep the garden clean by regularly raking up fallen leaves and clearing away any winter debris that has accumulated.

- If the ground isn't frozen or too wet, start preparing the soil for spring. Dig over empty beds and spread with a layer of organic matter, such as well-rotted manure or compost.

- Start to clean up your perennials. Bear in mind that some stems and flower heads, even when dead, can look beautiful.

- If you haven't done so already, cover delicate plants with fleece to protect them from wind and frost damage.

- Use black plastic or fleece tunnels to warm soil in the vegetable garden.

- Net any young vegetables to protect from birds.

LAWN CARE

- Frost can make grass brittle and easily damaged, so keep off the lawn as much as possible during the winter months.

TREES, SHRUBS, AND CLIMBERS

- If the ground isn't frozen, plant bare-root trees, shrubs, roses, and fruit bushes.
- Make sure that trees and shrubs are staked to stop them from being buffeted by strong winds and that climbers and wall shrubs are securely tied on to their supports.
- If boughs of trees and shrubs are straining under snow, brush it off to prevent them from getting damaged.
- Renovation-prune woody plants that are overgrown or congested by gradually cutting out unwanted growth.
- Prune standard apple and pear trees.
- Prune wisteria, cutting back to at least two buds, and remove any dead or damaged growth. Prune winter-flowering shrubs that have finished flowering, too.
- You can move established deciduous trees and shrubs now. Dig up as much of the root ball as possible and replant immediately. Water well and stake if necessary.

OTHER JOBS

- Make use of time indoors to plan what you want to grow during the year ahead and order in seeds.
- Wash and disinfect seed trays, pots, and tools.
- Clear dead plants from containers. Replant if temperatures allow, top up the soil, and finish with a layer of mulch.
- If you have a greenhouse, you can start sowing seeds.
- Remember to feed birds and other wildlife.

1 The yellow flowers of winter aconite (*Eranthis hyemalis*) are a cheerful sight early in the year.

2 After shedding its leaves, the white-stemmed bramble (*Rubus cockburnianus*) makes a dramatic thicket of ghostly white stems.

3 Not only do witch hazels (*Hamamelis*) look great at this time of year, but they smell fantastic, too.

FEBRUARY

Although it's the shortest month, February can sometimes feel really long. But remember, spring is (hopefully) just around the corner. Use this month to really start to think about the year ahead and—weather permitting—begin getting the garden in shape for the growing season.

QUICK CHECKLIST

- ○ **Divide bulbs "in the green."**
- ○ **Dig over any remaining empty beds you've not already dug over.**
- ○ **Improve any soil that is no longer frozen solid by digging in organic matter, such as well-rotted manure or compost.**
- ○ **Start sowing vegetable seeds and sprout potatoes.**
- ○ **Prune wisteria and summer-flowering climbers.**
- ○ **Begin weeding.**

Hellebore

IN YOUR BORDERS AND VEGETABLE GARDEN

- Continue to keep the garden clean by regularly raking up fallen leaves and clearing away winter debris.

- If the ground isn't frozen or too wet, start preparing the soil for spring. Dig over empty beds and spread with a layer of organic matter, such as well-rotted manure or compost.

- Continue cleaning up perennials and ornamental grasses by cutting off dead leaves, stems, and seed heads as necessary.

- If any perennials have outgrown their space, dig them out, divide them up, and replant them where they will have more room to grow.

- Begin to sow hardy annuals under cover, but don't plant too many seeds just yet as low light levels can result in weak plants. If any seedlings have their first true leaves, thin them out into pots. Handle seedlings by their first leaves rather than their stems.

- Divide bulbs such as snowdrops after they flower, but while they are still in leaf ("in the green"; see p211).

- Get ahead and sow some vegetables under cover in your greenhouse or a cold frame. Now is also the time to "chit" (sprout) potatoes.

- Net any young vegetables to protect from birds.

- Hoe or pull out small weeds, but if established, dig out the entire root.

1 When little else is flowering in the garden, the pretty nodding heads of snowdrops (*Galanthus*) are always a particularly welcome sight.

2 February brings carpets of early-flowering *Crocus tommasinianus*, with their small pretty purple flowers.

3 Wintersweet (*Chimonanthus praecox*), a vigorous shrub with sweet-smelling flowers, is great planted where the scent can be enjoyed in passing.

LAWN CARE

- Avoid walking on frosted lawns as it can damage the grass.

- In warmer areas, you might need to start mowing your lawn. For the first mow, set the blade high and trim off only the very top of the grass to avoid weakening it.

TREES, SHRUBS, AND CLIMBERS

- Plant bare-root trees, shrubs, roses, and fruit bushes if the weather is mild and the ground isn't frozen.

- Check that trees and shrubs are staked and that climbers and wall shrubs are securely tied in to prevent wind damage.

- Prune wisteria, cutting back to at least two buds, and remove any dead or damaged growth.

- Once the blooms on winter-flowering shrubs have finished flowering, prune them back ready for next year.

- Continue pruning standard apple and pear trees.

OTHER JOBS

- Clear dead plants from containers. Replant if temperatures allow, top up the soil, and add a layer of mulch.

MARCH

March is all about happy new beginnings and is the month many gardeners look forward to the most. Longer days and the promise of sunshine means that seeds germinate, bulbs bloom, and early shrubs begin to flower. It's the time when things really get going with a bang!

QUICK CHECKLIST

- ○ **Continue to weed.**
- ○ **Mulch beds and borders.**
- ○ **Cut back willow and dogwoods.**
- ○ **Last chance to plant bare-root trees, shrubs and roses.**
- ○ **Mow lawn as the weather improves.**
- ○ **Keep an eye out for slugs and snails.**

1

2

Daffodil

IN YOUR BORDERS AND VEGETABLE GARDEN

- There is still time to add a layer of mulch, such as well-rotted manure, to keep down weeds and retain moisture in the soil.

- Clean up perennials and ornamental grasses. Cut off dead leaves, stems, and seed heads where necessary.

- Dig out any perennials that have outgrown their space, divide them up, and replant where they'll have more room to grow.

- Feed the soil in borders with well-rotted manure or compost.

- Remove dead flowers from spring bulbs (but not the leaves) so they don't put all their energy into creating seeds.

- Continue sowing vegetables under cover.

- Net early crops to protect them from birds.

- Plant first early potatoes at the end of the month.

- Once weeds start to grow, the weather should be warm enough for you to sow seeds outside. Keep on top of weeds by hoeing them as they appear.

LAWN CARE

- The lawn can be mowed more often, but only on mild, dry days. You can now lower the mower blade, but never cut off more than the top one-third of the grass. Remember to spruce up the edges with an edging tool.

- This is the time to give the lawn a bit of extra attention. Toward the end of the month, feed it with fertilizer to increase vigor. Reseed any bare patches.

- Consider scarifying and aerating your lawn. This will help pull out thatch (dead moss and grass) and aid water filtration. Aeration is a good way to help water, air, and nutrients penetrate grass roots.

- If the weather is warm enough, you can lay new sod now.

TREES, SHRUBS, AND CLIMBERS

- Once the blooms on winter-flowering shrubs have finished flowering, prune them back ready for next year.

- This is the last chance to plant out bare-root trees, shrubs, and roses, so get these in the ground or into pots now.

- Finish winter pruning of standard apple and pear trees.

- Cut back willow and dogwoods to encourage bright new stem growth.

OTHER JOBS

- Dig out any trees that have self-seeded. You'll often find them beneath hedges where birds sit.

- Spring-clean garden furniture, all set for summer.

- Keep slugs and snails off plants. To deter them, I tend to use wool pellets in pots and containers, and nematodes in the borders.

1 **Weather permitting,** you can start laying new sod in spring.

2 **The early flowering** *Iris reticulata* **'George',** with its deep velvety purple flowers, is perfect for adding color to early spring days.

3 **The wild tulip,** *Tulipa sylvestris*, is one of my favorite native flowers, which I've naturalized under hedges and planted in my orchard.

APRIL

As new leaves and early flowers start to fill the garden, even fair-weather gardeners tend to get stuck in this month. It's great that there are more daylight hours because there's lots to do, particularly if you want to grow vegetables. However, there's still a chance of a late frost, so be ready to protect delicate blossom.

QUICK CHECKLIST

- ○ **Dig up and divide perennials.**
- ○ **Last chance to cut back perennials and ornamental grasses.**
- ○ **Put in plant supports.**
- ○ **Sow hardy annuals under cover.**
- ○ **Improve your lawn by scarifying and aerating it.**
- ○ **Keep on top of weeds.**

IN YOUR BORDERS AND VEGETABLE GARDEN

- Continue to feed the soil by digging in organic material like well-rotted manure and compost.

- This is your last opportunity to clean up perennials by removing dead material before they start growing.

- Continue to keep an eye out for perennial plants becoming congested in beds. Dig them out, divide them up, and replant where they'll have more room to grow.

- Plant tender, summer-flowering bulbs and tubers at the back end of the month.

- Sow seeds of hardy annuals under cover. As soon as seedlings have their first true leaves, select the onces that look vigorous and thin them out into pots. Handle seedlings by their first leaves rather than their stems.

- Put in stakes and plant supports for taller perennials.

- Sow vegetable seeds indoors and outdoors. You can also plant out crops, protecting tender plants with cloches or fleece, and plant midseason and main-crop potatoes.

- Net early crops to keep off birds.

- Top up raised beds with compost and good-quality topsoil.

- Feed plants with a balanced slow-release fertilizer.

- Keep pulling up, hoeing, and digging out weeds.

Spurge

LAWN CARE

- Lower the mower blade and cut grass shorter. Never cut off more than the top one-third.

- Fertilize the lawn to increase vigor, and treat it for moss and other weeds.

- Consider scarifying and aerating your lawn. This will pull out thatch (dead moss and grass) and help water, air, and nutrients penetrate grass roots.

- Continue to sow or lay new sod, keeping it well watered.

TREES, SHRUBS, AND CLIMBERS

- Prune late-summer flowering clematis. Cut plants back to knee height, above a group of flower buds. Feed and mulch with a general-purpose fertilizer.

- Prune evergreen shrubs, cutting out dead, damaged or diseased wood.

- Plant new evergreen trees and shrubs. This is also a good time to move established ones.

- Feed roses with manure, compost, or a balanced fertilizer.

OTHER JOBS

- Control diseases and pests, especially slugs and snails. To deter them, I tend to use wool pellets in pots and containers and nematodes in the borders.

- Top-dress pots and containers with some fresh compost, or replace the top 2in (5cm) if they are already full.

- Clean pavement. Pull up weeds, sweep, then scrub with a stiff brush and soapy water. If really dirty, use a diluted bleach solution (keeping away from plants). Rinse with clean water.

1 **Amelanchier lamarkii** is a really lovely, hardworking tree that has great year-round interest. The spring blossom is particularly beautiful.

2 **The candelabra primula**, *Primula japonica* 'Miller's Crimson', adds a lovely splash of color to the garden in spring.

3 **If the weather is warm enough**, sow annual seeds outdoors.

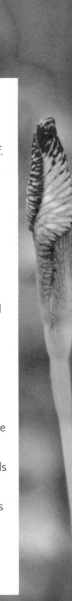

MAY

The garden really gets into full swing now. With warmer temperatures and longer days, hopefully you're thinking of having your first barbecue. If you really get on top of things this month, then the rest of the year will start to fall into place. Many spring bulbs will be over, but there's lots of stuff you can get lined up to follow.

QUICK CHECKLIST

- ○ **Water the garden during dry spells.**
- ○ **Plant out summer displays.**
- ○ **Sow fast-growing and late-flowering annuals.**
- ○ **Harden off tender seedlings.**
- ○ **Tie in climbing and rambling roses.**
- ○ **Watch out for late frosts.**

Wisteria

IN YOUR BORDERS AND VEGETABLE GARDEN

- Weed the garden as often as needed. A quick hoe around the garden on a warm, dry day is a good way to kill them off.

- Cut off the dead flower heads of spring bulbs, but leave the foliage to turn yellow.

- As growth takes off, keep an eye on congested beds. Continue to dig up and divide perennials to create new plants. Spring bulbs, such as daffodils, can also be divided.

- Thin hardy annuals to give them more growing space.

- Stagger the flowering season by giving some late-flowering herbaceous perennials a "Chelsea Chop": cut back one-third of the stems by one-third and another third by two-thirds.

- Plant out summer bedding displays and hardy annuals in beds and containers at the end of the month.

- Put in stakes and plant supports for taller perennials.

- Harden off tender seedlings, including half-hardy annuals, before planting out. Be prepared to protect them with fleece or cloches if there's a late frost.

- Start sowing culinary herbs. Runner beans and squash seeds can be sown directly outside into prepared beds.

- Water plants regularly during dry spells, ensuring containers and anything newly planted doesn't dry out.

- Earth up the soil around potatoes as they grow.

LAWN CARE

- Mow the lawn weekly. Never cut off more than the top one-third of the grass.

- Consider scarifying and aerating your lawn. This will pull out thatch (dead moss and grass) and help water, air, and nutrients penetrate grass roots.

- Continue to sow or lay new sod, keeping it well watered.

TREES, SHRUBS, AND CLIMBERS

- Tie in climbing and rambling roses to supports. Train stems horizontally to produce more side shoots and flowers.

- Prune spring-flowering clematis after it has flowered.

- Give trees, hedges, and other shrubs a boost with a slow-release mineral-rich fertilizer.

- Lightly clip evergreen hedges, but watch out for birds' nests.

OTHER JOBS

- Control pests and diseases, especially slugs and snails, leaf-rolling sawfly, and blackspot on roses.

- Top-dress containers with fresh compost and feed with a balanced liquid fertilizer every 2–3 weeks.

- Open the greenhouse on warm days.

1 *Iris* **'Tropic Night'** is one of my favorite plants and looks great with geums and geraniums.

2 This deep red *Aquilegia vulgaris* 'Ruby Port' is perfect for a cottage-garden feel.

3 Putting in plant supports early means they become hidden as the plants grow.

JUNE

This month sees the longest day of the year, and all the extra light produces an abundance of flowers. Keep weeding, filling gaps with tender and annual plants to ensure that the show continues throughout summer, and be sure to spend as much time as you can outdoors simply enjoying it all.

QUICK CHECKLIST

○ Tie in climbing and rambling roses.

○ Mow lawns regularly.

○ Neaten hedges.

○ Control pests and diseases.

○ Cut back early flowering perennials.

○ Be wise with water use.

Foxglove

IN YOUR BORDERS AND VEGETABLE GARDEN

- Continue hoeing off or digging up weeds as often as needed.

- Ensure containers and anything newly planted doesn't dry out. Be wise with your water use. Mulching with compost helps keep moisture in the soil and can provide nutrients, too.

- Cut back early flowering perennials, such as hardy geraniums, to encourage fresh foliage and possibly more flowers.

- Deadhead spent flower heads to encourage flowering, unless you're collecting seeds.

- If you have any gaps, plant out summer bedding to add color to planting schemes.

- Tender seedlings grown under cover can be planted out.

- Protect any developing fruit on smaller plants with netting.

- Start sowing your biennial seeds directly into the ground.

- Make sure plants are tied to supports or staked if needed.

LAWN CARE

- Mow lawns regularly. Never cut off more than the top one-third of the grass.

- Don't forget to check the lawn for weeds.

- Make sure new areas of lawn that were sodded or sown in spring don't dry out during hot spells.

TREES, SHRUBS, AND CLIMBERS

- Some early flowering shrubs, such as lilac, spirea, broom, and deutzia, can be pruned now.

- Keep hedges trimmed and neat, but watch out for birds' nests. Lightly clip evergreen hedges.

OTHER JOBS

- Control diseases and pests, particularly slugs and snails. To deter them, I tend to use wool pellets in pots and containers and nematodes in the borders.

- Watch out for aphids, especially on the back of leaves. Rub them off or spray them with soap-based insecticide.

- In hot spells, water containers and hanging baskets daily and feed with a liquid fertilizer every 2–4 weeks.

- Be wise with water in drought-affected areas, using rainwater or gray water if possible.

> *If you have any gaps, plant out summer bedding to add more color to borders.*

1 Deadhead plants regularly to keep borders looking neat and to help plants conserve energy.

2 Angelica archangelica is an impressive biennial that grows to 6ft (2m) with large umbel flowers.

3 A great plant for pollinators, the wine color of *Cirsium rivale* 'Atropurpureum' mixes well with blue iris and pale orange geums.

JULY

I love this month! Borders are looking great, there's lots of lovely scent, and there are plenty of vegetables to harvest. Make sure you water plants in dry weather, but, most important of all, go out in the long evenings after a day's work and enjoy how lovely it all is.

QUICK CHECKLIST

- ○ **Water regularly, particularly during hot, dry spells.**
- ○ **Keep deadheading regularly.**
- ○ **Continue to plant out vegetable seedlings.**
- ○ **Feed roses.**
- ○ **Cut back wisteria shoots.**
- ○ **Prepare to sow new lawns.**

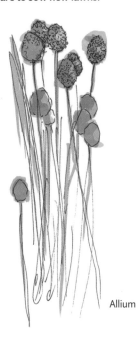

Allium

1

2

IN YOUR BORDERS AND VEGETABLE GARDEN

- Water plants regularly during dry spells, ensuring containers and anything newly planted doesn't dry out. Be wise with your water use. Mulching around plants with compost helps keep moisture in the soil and can provide nutrients, too.

- Keep deadheading to encourage more flowering, unless you're planning to collect seeds.

- Cut back perennials that have finished flowering to encourage new foliage and flowers.

- There's still plenty to sow directly outside this month, including biennials and vegetable seeds.

- Continue to plant vegetable seedlings that were grown in trays or pots under cover directly into the ground.

- Keep checking whether any plants, including climbing vegetables, need tying in or staking.

LAWN CARE

- Mow lawns regularly. Never cut off more than the top one-third of the grass.

- If you plan to sow a new lawn this year, start preparing the ground now.

TREES, SHRUBS, AND CLIMBERS

- Prune early flowering shrubs that have finished flowering, such as lilac, spirea, broom, and deutzia.

- Feed roses to encourage a second flush of flowers and mulch around the base to retain moisture.

- Prune climbers, including spring-flowering clematis.

- Summer prune fruit trees, such as apple and pear trees, and thin out fruits.

- Cut back long, wispy wisteria shoots to seven buds from the main stems.

- Ensure climbing plants are tied securely to their supports.

- Trim and neaten hedges, but watch out for birds' nests.

OTHER JOBS

- Take measures to control pests and diseases, including slugs and snails. To deter them, I tend to use wool pellets in pots and containers and nematodes in the borders.

- In hot spells, water containers and hanging baskets daily and feed with a liquid fertilizer every 2–4 weeks. Be wise with water in drought-affected areas, using gray water or rainwater if possible.

- Keep ponds and water features topped up and clean.

- If you're going on vacation, arrange for a friend to keep an eye on the garden and water plants.

- Take pictures of your garden and make a note of anything you want to change or move later in the year.

1 Now's a good time to cut back the long whippy growth of wisteria**.**

2 *Potentilla* **'Monarch's Velvet'** is very long-flowering and great for a sunny border.

3 The large blue thistle heads of *Eryngium bourgatii* add a spiky drama to planting schemes.

AUGUST

The summer holiday season is the time to really make the most of your garden as it's (supposedly) the hottest month. It's a great time to be outdoors cooking and eating with family and friends. Keep the garden in good shape, water plants on hot, dry days, and feed containers.

QUICK CHECKLIST

- ○ Water plants in hot, dry spells.
- ○ Keep deadheading regularly.
- ○ Order spring-flowering bulbs.
- ○ Prune climbing and rambling roses.
- ○ Feed perennials in beds and borders.
- ○ Sow quick-growing salad crops.

Selinum

IN YOUR BORDERS AND VEGETABLE GARDEN

- If there's a drought, focus on watering new plants in the evening or early morning. Be wise with your water use. Mulching around plants with compost helps keep moisture in the soil and can provide nutrients, too.

- Keep deadheading to encourage more flowering and keep the garden looking tidy, unless you want to collect seeds.

- Feed tired-looking perennials with a liquid feed.

- Cut back foliage and stems of plants that have died back. Leave any nice-looking flower and seed heads in place to provide interest during fall and winter months.

- Place stakes around the garden with labels showing what you plan to dig and divide later in the year.

- In the vegetable garden, keep sowing quick salad crops such as lettuce, radish, and arugula.

LAWN CARE

- If you haven't done so already, summer prune trained fruit trees now to maintain a good shape and let light in.

- If you plan to sow a new lawn this year, start preparing the ground now if you haven't already done so.

TREES, SHRUBS, AND CLIMBERS

- Prune climbing and rambling roses that do not repeat flower or produce hips, cutting back dead, diseased, or damaged branches to the ground or to a healthy bud.

- After flowering is over, prune climbing shrubs and cut back long, wispy wisteria shoots to seven buds from the main stems.

- If you haven't done so already, summer prune trained fruit trees now to maintain a good shape and let light in.

- Clear away leaves that have fallen beneath roses and burn them to prevent proliferation of black spot.

- Trim and neaten hedges, but watch out for birds' nests.

OTHER JOBS

- Control pests and diseases, especially slugs and snails. To deter them, I tend to use wool pellets in pots and containers and nematodes in the borders.

- Water plants in containers and hanging baskets regularly and feed them with a liquid fertilizer every 2–4 weeks. Be wise with water in drought-affected areas, using gray water or rainwater if possible.

- If you're going on vacation, arrange for a friend to keep an eye on the garden and water plants.

- Order spring-flowering bulbs to plant in the fall.

- Take pictures of your garden and make a note of anything you want to change or move later in the year.

1 Now's the time to harvest crops, such as tomatoes. It's a great way to get kids involved in the garden.

2 *Rosa* 'Gertrude Jekyll' is my all-time favorite rose and flowers all summer long.

3 The purples and pinks of *Sanguisorba* and poppies combine well togeher.

SEPTEMBER

This is a joyful time of year with wonderful light,
leaves gradually changing color, and berries appearing.
The abundant harvest always makes
me think of my grandad, a keen vegetable grower.
Now is the perfect chance to begin tidying up the
garden before winter starts to set in.

QUICK CHECKLIST

○ **Collect and save seeds.**

○ **Start planting spring bulbs.**

○ **Put new perennial plants, trees, and shrubs in the ground.**

○ **Prepare the vegetable garden for winter.**

IN YOUR BORDERS AND VEGETABLE GARDEN

- This is a key time to plant for the year ahead. Get spring-flowering bulbs into the soil, but hold back on tulips. Biennials sown inside can now be planted out. It's also a good time to plant new perennials and direct sow hardy annuals. Keep watering newly planted plants.

- Deadhead summer blooms regularly unless you want to gather seeds from them. Store any collected seeds in paper bags in a cool dry place.

- Cut down perennial plants that have died back, except for flower and seed heads that will provide winter interest.

- Start to divide herbaceous perennials and replant them where they will have room to grow.

- Prepare vegetable beds for winter: clear away old vegetation, remove supports, dig over beds, and sow vegetables that will overwinter.

Coneflower

LAWN CARE

- The grass won't grow so quickly now, so you can mow less. Raise mower blades slightly to avoid weakening the grass.

- Consider scarifying and aerating your lawn. This will pull out thatch (dead moss and grass), and help water, air, and nutrients penetrate grass roots. Apply a fall lawn feed after scarifying and aerating to prepare your lawn for winter.

- This is a good time to sow lawn seed or lay down new sod.

TREES, SHRUBS, AND CLIMBERS

- Prune climbing and rambling roses that do not repeat flower or produce hips, cutting back dead, diseased, or damaged branches to the ground or to a healthy bud.

- Plant new trees and shrubs.

- Prune late-summer-flowering shrubs and give evergreen hedges a final clean up before winter.

- Clear away leaves that have fallen beneath roses and burn them to prevent proliferation of black spot.

OTHER JOBS

- Control pests and diseases, keeping an eye out for powdery mildew or rust.

- If you garden on heavy clay, this is a good time to dig in organic matter to improve soil structure.

- It can still be a hot, dry month so continue to be wise with your water use in drought-affected areas, using gray water or rainwater if possible.

- Clean out your greenhouse ready for fall use.

- Take photos to keep a record of your September garden and make a note of anything you want to change later.

1 Collect seed heads of your favorite herbaceous perennials.

2 _Verbena bonariensis_ produces a haze of purple flowers that are an absolute magnet for bees and butterflies. Its strong wiry stem can reach 6ft (1.8m).

3 The nodding yellow flower heads of _Clematis tangutica_ are followed by fluffy seed heads.

OCTOBER

This is an elegant month with lots of beauty in fall color, and I love seeing the garden's structure reveal itself again. With shorter days, we tend to retreat indoors, but there's still lots that can be done in the garden. Now's the time to prepare the for winter and plant spring-flowering bulbs if you've not already done so.

QUICK CHECKLIST

○ **Dig up and divide congested clumps of perennials.**

○ **Continue planting spring bulbs.**

○ **Keep collecting seeds from your favorite plants.**

○ **Mulch beds and borders.**

○ **Last chance to sow grass seed.**

○ **Clean and store garden furniture.**

IN YOUR BORDERS AND VEGETABLE GARDEN

- Cut back perennial plants that have begun to die back, leaving in place any attractive flower or seed heads that will add winter structure.

- Collect seeds and store them in a cool, dry place.

- Split overcrowded clumps of perennials by digging them up and dividing plants.

- Continue to plant spring-flowering bulbs (apart from tulips).

- Put down a heavy mulch to protect tender perennials or cut back their stems and lift and store the plants somewhere cool and dry.

- Mulch over bare soil in beds and borders. Use homemade compost, leaf mold, or green waste and spread it at least 1½in (4cm) deep.

- If you grow vegetables, clear away old vegetation, remove supports, and dig over beds, working in well-rotted manure. Now is the time to plant onions and garlic, too.

- Divide established rhubarb crowns to create new plants.

- Keep watering newly planted plants, if necessary.

Katsura tree

1 *Rudbeckia fulgida* **var.** *deamii* **(black-eyed Susan)** produces masses of yellow daisy flowers with a dark brown central cone.

2 *Agastache* **'Black Adder'** is great for adding a bit of height to a border. Butterflies love it too.

3 The fall colors of the Japanese maple, *Acer palmatum*, are exceptionally good.

LAWN CARE

- You can mow less now the grass growth has slowed. Raise the blades to avoid weakening the grass.

- This is the last month you can scarify and aerate your lawn to remove dead material and improve root health. It's also your last chance to sow grass seed. Sod can be laid if it's not too cold or wet.

TREES, SHRUBS, AND CLIMBERS

- Prune climbing and rambling roses that do not repeat flower or produce hips, cutting back dead, diseased, or damaged branches to the ground or to a healthy bud.

- Bare-root trees, shrubs, and roses can be planted now, as well as container trees, shrubs, and climbers.

- Clear away and burn leaves that have fallen beneath roses to prevent the spread of black spot.

OTHER JOBS

- Clean outdoor furniture, cover, and store somewhere dry.

- Clean pavement. Pull up weeds, sweep, then scrub with a stiff brush and soapy water. If really dirty, use a diluted bleach solution (keeping away from plants). Rinse with clean water.

- Take photos of plants before they die back as a useful reminder of where they are when planning for next year.

NOVEMBER

I love the light in November, and, with summer displays pretty much over, the garden starts to look romantically wintry. Try not to let early frosts catch you unawares by being ready to protect tender plants if need be, and give yourself a head start on next year by making preparations now.

QUICK CHECKLIST

- ○ **Finish planting spring bulbs and start putting in tulips.**
- ○ **Plant bare-root trees and shrubs.**
- ○ **Dig over empty beds and spread manure on them.**
- ○ **Protect containers from frost.**
- ○ **Rake up fallen leaves.**
- ○ **Give lawns a final cut.**

IN YOUR BORDERS AND VEGETABLE GARDEN

- Put in the last of your spring-flowering bulbs and, once temperatures drop, start planting tulips.
- Continue to split up overcrowded clumps of perennials by digging the plants up and dividing them.
- If you haven't already done so, put down a heavy mulch to protect tender perennials, or cut back their stems and lift and store the plants somewhere cool and dry.
- Cut back perennial plants that have started to die back, leaving any attractive flower or seed heads in place for winter structure.
- Continue to mulch bare soil, spreading it with well-rotted manure, homemade compost, leaf mold, or green waste.
- Clean up leaves as they fall, especially on lawns and in ponds. Store them for leaf mold if you have space.
- Give borders a final clean up and remove any remaining weeds before winter.
- Clear away crops that have finished in the vegetable garden, tidy up supports, and dig over beds, working in well-rotted manure.

Mahonia

1 **The long shiny red hips** of *Rosa* 'Doncasterii' look a bit like chile peppers and make a great winter display.

2 **Protect tender plants** by covering with layers of horticultural fleece secured with string.

3 **Crimson glory vine**, *Vitis coignetiae*, is fast growing, with spectacular fall colors.

LAWN CARE

- Give the lawn its final cut of the year, then clean and put the mower away. Keep an eye out for any areas where water lies over the winter; this will help guide improvement next year.

TREES, SHRUBS, AND CLIMBERS

- Prune climbing and rambling roses that do not repeat flower or produce hips, cutting back dead, diseased, or damaged branches to the ground or to a healthy bud. Prune roses in borders, too. Reducing their height will help prevent wind rock during winter.

- Bare-root trees, shrubs, and roses can be planted now. Container trees and shrubs can still be planted, too.

- You can move established deciduous trees and shrubs as they become dormant. Dig up as much of the root ball as possible and replant immediately. Water well and don't forget to stake if necessary.

OTHER JOBS

- Protect containers from frost by wrapping them in fleece or hessian. Alternatively, bring them indoors.

- Check your tools and do any necessary maintenance.

DECEMBER

Shorter, chilly days make December the ideal time to stay indoors by the fire with a stack of seed catalogs. But make sure you still walk around the garden area, too. With everything laid bare, it's a great time to see whether you need to add more winter interest or structure.

QUICK CHECKLIST

- ○ Clean up leaves and debris.
- ○ Brush snow from tree boughs.
- ○ Plant bare-root trees, shrubs, and roses.
- ○ Dig over and prepare vegetable beds.
- ○ Start planning for next year.
- ○ Order seeds.

IN YOUR BORDERS AND VEGETABLE GARDEN

- Regularly rake up and clear any leaves and debris in the garden, storing them for leaf mold if you have space.

- Tidy up your perennials if they become frosted. Bear in mind that some stems and flower heads, even when dead, can look beautiful.

- If there's a frost, look out for any border plants, bulbs, or shrubs that might have heaved up out of the ground as the soil froze and swelled upward. If so, gently firm them back in.

- As long as there isn't a frost, you can dig over vegetable beds and spread with well-rotted manure or compost.

Witch hazel

3

> *Now is the time to observe the bones of your garden's structure.*

LAWN CARE

- Frost can make grass brittle and easily damaged so keep off the lawn as much as possible during the winter months.

TREES, SHRUBS, AND CLIMBERS

- Make sure climbers, shrubs, and trees are securely tied to their supports and stakes.

- Prune standard apple and pear trees.

- Shake snow off trees and shrubs to prevent damage.

- As long as the ground isn't frozen, you can plant bare-root shrubs, trees, and roses into the ground now.

- Established deciduous trees and shrubs can be moved now that they are dormant. Dig up as much of the root ball as possible and replant immediately. Water well and stake if necessary.

OTHER JOBS

- If you haven't done so already, protect containers from frost by wrapping them in fleece or hessian. Alternatively, bring them indoors.

- Begin planning now for next year's growing season and start ordering in fresh seeds.

1 Yew (*Taxus baccata*) makes a lovely dense hedge that can be clipped into all sorts of shapes to suit any style of garden. Note all parts are toxic to humans.

2 The glossy, dark green, arrow-shaped leaves of *Arum italicum* subsp. *italicum* 'Marmoratum' add great winter interest.

3 Winter-flowering clematis have pretty flowers that brighten the garden during the colder months.

USEFUL
INFORMATION

MEASURING YOUR SITE

Before you start any building work, you may have to calculate the slope of a terrace, figure out how to step fence panels along a slope, or decide how many steps your garden area can comfortably accommodate. Follow the advice below on ways to help you plan your project.

Note that in this book I've included both imperial and metric measurements. Whatever system you use, make sure you're consistent and use either all imperial or all metric measurements for individual projects. Here, I've used imperial units only for some calculations to keep things simple.

USEFUL INFORMATION

SETTING A **LEVEL** FOR **RAINWATER TO RUN OFF**

Putting a slight slope on your paving helps rainwater run off easily so you don't end up with puddles or damp problems with your house. When you buy materials, information about the recommended gradient will be available from your supplier. The gradient will read as a ratio, usually between 1:40 and 1:100, which is the fall or drop you need on your paving.

If you're building a terrace next to a house, ideally the paving should sit 6in (15cm) below the dampproof course. (This rule doesn't have to apply all the time, but do your research before breaking it.) Look for a black liner between the bricks near the foot of the brickwork on your house—this is the damp course.

If the drop of your terrace is a ratio of 1:80, it means that for every 80 units of distance, the level should fall by 1 unit. So, for example, if your terrace extends 6½ft (2m), or 79 in (200cm), from the house, you divide 79 by 80, which is 1in (2.5cm). So the terrace should sit 1in lower on the side farthest from the house. Once you've excavated the area to be paved, the next step is to mark out the slope using these measurements.

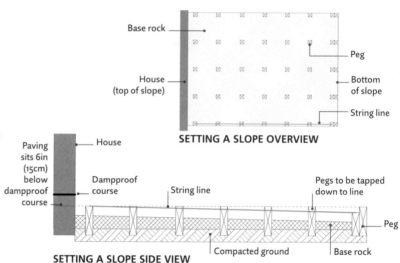

SETTING A SLOPE OVERVIEW

SETTING A SLOPE SIDE VIEW

1 Starting at the top of your slope and working away from the house, put in pegs in straight lines across the site. Put all pegs in to the finished paving level. Check your levels across the site with a spirit level. Make sure you knock in a line of pegs at the bottom of the slope, keeping them level.

2 Next to the pegs at the bottom, knock in pegs at the lowest level of the slope to your required drop.

3 Fix a string line at right angles to the house from a peg at the top of the slope to one of the lower pegs at the bottom. Make sure it is tight. Working back to the top of the slope, tap down the other pegs along the line so they are flush with the line.

4 Then tap in the other pegs across the site so that, at each point on the slope, they are flush with the pegs along the string line. Check your levels across the site with a spirit level and a straight plank of wood. These pegs are a good reference point for checking levels as you lay your paving.

5 Use a pencil and set square to mark the base rock level onto your pegs. Remember to measure from the top and allow for the depth of your paving material and mortar bed.

6 After you've put in the base rock to the level you marked on the pegs, tamp down the site and use a plate compactor to firm everything down. Now you're ready to lay your paving.

CHECKING THE LEVELS OF POSTS

To check that posts are horizontally level, lay a straight plank of wood across the top of the posts and put your spirit level on top. Adjust the levels of the posts as necessary. If your posts need to drop down an uneven slope, put a block of wood cut to the required drop on the post that is downslope (see Post B in the diagram far right). Then check the levels with a plank of wood and a spirit level as before. You can use Post B to check the level on the next post downslope, and so on.

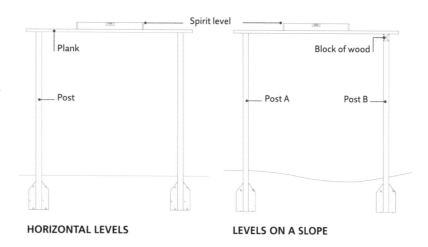

HORIZONTAL LEVELS

LEVELS ON A SLOPE

STEPPED FENCING ALONG A SLOPE

If you're erecting a fence on a slope with preassembled panels, you'll have to step the posts to follow the slope. In other words, the line of the fence follows the slope like a stairway. Usually, posts need to be at least 2ft (60cm) longer than the height of the panel, but you may need to have longer posts with a stepped fence to accommodate big changes in height.

Calculate the length of the fence and the drop in height, and then spread the drop evenly between the posts.

It's a good idea to create a scale drawing before you start erecting any fencing, as you'll need to decide how to step the panels to follow the slope. If the slope is steep, you'll probably have gaps beneath parts of the panels big enough for children or pets to get through. If that's the case, you may need to infill with gravel boards.

Example

Fencing is to be erected along a sloped boundary of 98ft with a drop of 39in. The width of one fence panel and one post is 6ft (see p246). You divide the length of the boundary by the panel and post width to calculate how many panels you need: 98/6 = 16.3, rounded down to 16.

Then you divide the drop by the number of panels to calculate the drop per panel: 39/16 = 2⅜in.

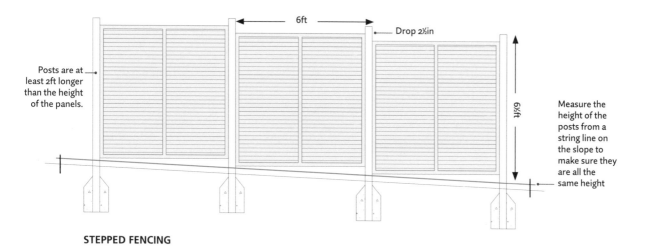

Posts are at least 2ft longer than the height of the panels.

Measure the height of the posts from a string line on the slope to make sure they are all the same height

STEPPED FENCING

CALCULATIONS FOR STEPS

When building steps, you need to know the vertical and horizontal measurements. The height at which your steps start and finish is the vertical measurement, and the horizontal measurement is how far the steps extend into the garden area. You need to measure from fixed points at the top and bottom of the slope.

There's a simple way to figure out how many steps you can accommodate comfortably. First, divide the vertical height by the riser height to calculate the number of steps. Then you can divide the horizontal measurement by the number of steps to find the tread depth. Generally, the standard riser is about 6in (15cm) high, and the tread should not be less than 12in (30cm) deep, but you can be more generous if you have the space. It's vital to have all your risers the same height; otherwise, they become a serious trip hazard.

There can be lots of variables, so take your time to figure out the best option for you. You might have a confined space, or you might be able to build the steps as far into the garden as you want. Sometimes it's possible to make adjustments to your vertical height by digging out or raising the retaining wall slightly, in which case you may be able to adjust your calculations slightly. The materials you choose can also affect your calculations, and you may need to cut materials to fit.

BRICK STEPS

The height of a standard brick is 2⅝in, so along with ⅜in of mortar, the height of two bricks measures ⅜in + 2⅝in + ⅜in + 2⅝in, which is 6in, the perfect height for a riser.

Example 1

The vertical measurement is 17in. Divide the vertical measurement by the riser height to find the number of steps:

17 / 6 = 2.8 steps.

Since this is close to being 3 steps, you could reduce the riser height slightly. To do this, divide 17 by 3, which is 5⅔in. So you could make the risers slightly lower than the standard size to accommodate the steps.

Example 2

The vertical measurement is 19½in:

19½ / 6 = 3¼ steps, rounded down to 3.

You need to increase your riser height slightly to accommodate the vertical measurement. First, multiply the riser height by 3:

3 x 6in = 18in

Then subtract that number from your vertical measurement:

19½ – 18 = 1½in to accommodate over 3 steps

1½in / 3 = ½in per step. Add that to the riser height:

6in + ½in = 6½in.

So the riser measurement for each of the 3 steps is 6½in. To check that this number is correct, multiply the number of steps by the riser height:

6½in x 3 = 19½in.

Next, you have to figure out the depth of the treads. You can go deeper but not less than 12in (30cm), but be mindful that the depth of your tread will have a direct impact on how far the steps extend into the garden. Where things become tricky is if you have a confined space, since this will limit how deep your treads can be.

Example 3

The vertical measurement is 17in. You've calculated that you can have 3 steps with a riser height of 5⅔in.

The horizontal measurement, or the maximum distance you can extend the steps into the garden, is 47in. Divide this by 3 steps to find the depth of the treads:

47in / 3 = 15⅔in.

Example 4

The vertical measurement is 17in. If you can extend the steps into the garden by only 27in, you will need to recalculate to fit the steps comfortably: 27 / 3 = 9in. This is less than the standard tread depth of 12in and so is too shallow. You will need to recalculate the number of steps you have. Repeat the calculations to see how things would work if you reduced the steps to 2. You will also need to recalculate the new riser height:

17in / 2 = 8½in.

To calculate the new size tread:

27in / 2 = 13½in.

So the best option with a vertical measurement of 17in and a horizontal measurement of 27in is to have 2 steps with risers 8½in high and a tread depth of 13½in.

CALCULATING QUANTITIES

It's important to have an accurate measure of the quantities of materials you need before you start building. Follow the advice below on ways to help you measure up for quantities using your scale plan.

Note that the calculations in this section are all approximate. Speak to your building materials supplier to get advice on quantities for specific materials, or check all your measurements on an online calculator if you're ordering online.

CHECKING A **RIGHT ANGLE**

One of the challenges of building is making sure you have corners absolutely square. You can create an accurate 90° angle using a method known as the **3-4-5 triangle**.

It is based on Greek mathematician Pythagoras's finding that the sides on a right-angled triangle are in the ratio 3:4:5. One side measures 3 units, the next 4, and the longest side measures 5 units. You can use line and pins to form a triangle that has multiples of 3:4:5 (such as 3:4:5in or 30:40:50cm) and be confident that you have a right angle between the two shorter sides.

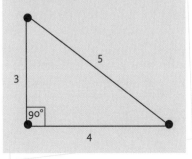

HOW MUCH **BASE ROCK?**

The composition of base rock varies enormously, and therefore the coverage will vary, too. Since base rock involves depth as well as area, you need to work in cubic units to calculate volume. In general, volume = width x height x depth.

Example

The area of a terrace is 10ft x 13ft = 130 sq ft (ft²). To calculate the volume of base rock needed to lay it to a depth of 4in (⅓ft), multiply the area by the depth: 130 x ⅓ = 43 cubic ft (ft³).

HOW MUCH **MORTAR?**

I tend to allow 8 to 10 bags of cement for 3ft² when constructing, which is also a fairly accurate number for general landscaping. To calculate the volume of sand and cement you need to make mortar, divide the total volume by the sum of the ratio numbers. Then multiply that number according to the ratios needed.

Example 1

The area of a terrace is 10ft x 13ft = 130ft². To lay mortar to a depth of 1½in (⅛ft), multiply area by depth:
130 x ⅛ = 16ft³.

You then need to figure out the quantities of cement and sand, depending on the proportions of your mortar mix.
For a **6:1 mix** of sand to cement, divide the total volume by 7:
16 / 7 = 2ft³.
So 1 part cement is 2ft³, and 6 parts sand is 6 x 2 = 12ft³.

Example 2

For a **4:1 mix** of sand to cement, divide the total volume by 5:
16 / 5 = 3ft³.
So 1 part cement is 3ft³, and 4 parts sand is 4 x 3 = 12ft³.

HOW MUCH **GRAVEL?**

Gravel is available in many types, finishes, and sizes. It also comes in different sizes of bags. The smaller the bags, the higher the cost if you are covering large areas. Buying a loose tip load will be more cost effective.

I tend to use the same calculation for calculating quantities as when I'm buying base rock, which is area x depth. However, check with your supplier because coverage can vary.

Example

The area of a terrace is 10ft x 13ft = 130ft^2. To lay gravel to a depth of 1¼in ($\frac{1}{10}$ft), the volume of gravel needed is:
130 x $\frac{1}{10}$ = 13ft^3.

HOW MANY **FENCE PANELS** AND **POSTS?**

Check your scale plan. Measure the boundary where you require fencing and divide that by the width of your fence panels and posts. You need to allow one post per panel, apart from the first panel, which needs an extra post. For example, if you have 10 panels, you will need a total of 11 posts. Remember that the posts need to be at least 2ft longer than the height of the panels.

Example:

A boundary is 98½ft long.
1 fence panel measures 6ft wide.
1 fence post measures 4in (or ⅓ft) wide.
1 fence + 1 post = 6 + ⅓ = 6⅓ft.
The number of fence panels needed is 98½/ 6⅓ = 15½, rounded up to 16 panels and 17 posts.

Note that 16 fence panels + 17 posts = 16 x 6⅓ft + ⅓ft = 101½ft. So the last fence panel (ideally, the farthest from view) will need to be cut to fit.

HOW MANY **SLABS** OR **BRICKS?**

To calculate how many paving slabs or bricks you'll need for your terrace, use this simple calculation.

First, calculate the area of the paving (width x length) and convert ft^2 to in^2. Then figure out the area of 1 slab or brick in in^2.

Next, calculate how many slabs are needed per ft^2 by dividing 1ft^2 by the area of 1 slab or brick.

Then multiply the area of the terrace by the rate per square foot to give the total number of slabs needed, plus about 10% for wastage. Wastage may vary depending on pattern and material.

If you're not confident about doing the math accurately, use one of the many online paving calculators. Your supplier may be able to help with the calculations, too.

HOW MUCH **SOD** DO I NEED?

Sod pieces are sold in square feet, so figure out your areas in that unit.

Example

The area of lawn to be sodded is 10ft x 10ft = 100ft^2.
Add 10% to 20% for wastage (depending on the shape).
100 + 10 (10% wastage) = 110ft^2.
100 + 20 (20% wastage) = 120ft^2.

Example 1

The area of a terrace is 10ft x 13ft = 130ft^2.
The area of 1 slab is 24in x 17in = 408in^2.
1ft^2 = 144in^2.
The number of slabs needed for 1ft^2 is 144/408 = 0.35, rounded up to 0.5. The number of slabs for 130ft^2 is 130 x 0.5 = 65. Add 10% (6.5) for wastage:
65 + 6.5 = 71.5, rounded up to 72 slabs needed in total.

Example 2

The area of a pathway is 5ft x 33ft = 165ft^2.
The area of 1 brick is 8½in x 2½in = 21¼in^2.
1ft^2 = 144in^2.
The number of bricks needed for 1ft^2 is 144/ 21¼ = 6.78, rounded up to 7.
The number of bricks for 165ft^2 is 7 x 165 = 1,155
Add 10% (115) for wastage:
1,155 + 115 = 1,270 bricks needed in total.

Note When calculating how many bricks you need to build a 9-inch brick wall (see p172), remember that you'll need two courses of bricks. However, if the back is not visible, then you can substitute with cheaper bricks or concrete blocks.

HOW MUCH **GRASS SEED** DO I NEED?

Grass seed is sold by the pound, and you'll need to calculate the square footage of the area to be sown. Check the seeding rate for the sort of grass seed you plan to sow, as it can vary between ⅓lb and 10lb per 1,000ft^2, depending on the seed.

Example

The area to be seeded is 13ft x 20ft = 260ft^2.
If the seeding rate is 7lb per 1,000ft^2 (or 7/1,000 = 0.007lb per 1ft^2), the weight of seed needed for 260ft^2 is 0.007 x 260 = 1.8, rounded up to 2lb.

GLOSSARY

AGGREGATE
In the building industry, aggregate refers to a broad category of coarse particles used for construction and includes things such as gravel, crushed stone, and sand.

ANNUALS
Plants that perform their entire life cycle from seed to flower to seed within a single year. All roots, stems, and leaves of the plant die annually. Only the dormant seed bridges the gap between one generation to the next.

BIENNIALS
Plants that require two years to complete their life cycle. In the first season, the plant grows roots, leaves, and stems. During the second season, the stems grow longer, flowers develop and bloom, and then seeds are formed, after which the plant dies.

BINDING SURFACE
This is often used to help bed gravel when base rock is very hard and does not give. I usually use ballast.

BULBS
Plants that grow from fleshy "bulbs" planted underground. The term is sometimes used to describe not only true bulbs but also corms and rhizomes. Generally, after flowering in spring or summer, they lose their upper parts, including flowers, stem, and leaves, and survive the rest of the year by living off the nutrients and moisture retained in the plant's storage system during the growing season.

CLIMBERS
Plants that grow upward by attaching themselves to other plants or objects. Some support themselves using tendrils or suckers, while others will need to be tied onto a supporting frame such as a trellis.

CONCRETE
Concrete is a mixture of water, ballast, and cement. The mixes change depending on use. It is generally used for footings beneath walls but can also be used in different ornamental ways.

DECIDUOUS
This term refers to a tree or shrub that sheds its leaves, usually each fall, then grows new ones the following year, usually in spring. Before the leaves drop, they often change color, gradually falling until the tree's branches are bare.

EVERGREEN
Plants that retain their leaves all year round, shedding old ones and growing new ones throughout the year, so they never have bare branches.

FOOTING
A footing is a solid, durable foundation used beneath the base of a brick wall or steps to make sure everything remains stable. Normally, it is formed by digging a trench, filling it with concrete, tamping it down until it's firm and level, then letting it set for at least 24 hours, but normally longer.

GRANITE SETTS
Generally, these are small rectangular blocks of quarried stone made for paving roads and paths.

HALF-HARDY ANNUALS
These are annual plants that don't like being in frozen ground so are best planted in late spring when all danger of frost has passed. Alternatively, start them off indoors early in the year then plant them out later.

HALF-HARDY/TENDER PERENNIALS
Tender perennial plants that won't survive if they are left outdoors in frosty weather. Alternatively, you can leave them out and replace them with new plants each year.

HARD LANDSCAPING
Also called "hardscape," this term refers to the construction materials used in garden and landscape design, such as stone, wood, brick, concrete, steel, and gravel.

HARDY ANNUALS
Plants that can be sown directly into the ground either in fall or early spring. Unlike tender annuals, they can

withstand some frost, although they are best protected with fleece if it's particularly heavy.

HEDGING
A hedge is a line of closely planted shrubs or trees, evergreen or deciduous, that knit together to form a barrier and habitat.

MORTAR
This term is used to describe a mix of sand, cement, and water. It is generally used to bind masonry works (e.g., brick and stone) and lay paving.

NATURALIZE
If you see a plant labeled "good for naturalizing," it means that it will spread naturally and informally around your garden, coming up year after year. Certain types of daffodil and tulip look particularly good naturalized in rough grass.

PERENNIALS
Essentially, these are plants that persist for several growing seasons and generally die back each winter and regrow the following spring from the same root system. Some are short-lived while others can be long-lived. Perennials have nonwoody stems and bring interest, flowers, and structure to borders. Some are classed as evergreen, or nonherbaceous, and keep their leaves all year round. Herbaceous perennials, on the other hand, die back over winter. Their roots

survive below ground, and the plant regrows again in spring to perform the yearlong cycle once more.

SCALE
When designing a garden, we use scale to translate measurements taken from the actual landscape and reduce them to a workable size that we can fit on a piece of paper. This means we can then easily design the whole space, as well as use the measurements to figure out useful information, such as quantities needed. It is usually shown as a ratio, for example, 1:50 or 1:100.

SHRUBS
A woody plant that is normally smaller than a tree and has several main stems arising at or near the ground.

SHUTTERING
Planks of wood used to contain setting concrete or to support the sides of a trench.

SOD
Sod is basically grass and the layer of earth held together by roots beneath it. You can buy rolls of sod to lay a lawn.

SOFT LANDSCAPING
Soft landscaping, also known as "softscape," describes the plant material used in garden and landscape design, including trees, shrubs, perennials, grass, and bulbs.

SPECIES TULIP
These are tulips that occur naturally in the wild. Often they are smaller and have a more understated charm than bigger and bolder tulips that have been bred for ornamental purposes.

SUCKERING HABIT
Some plants spread by means of suckers, which are vigorous growths of new stems and root systems that emerge from the existing roots or the lower stem of a plant. Unless you want the plant to spread, it's best to remove suckers because they can sap a plant's energy.

TAMP
To compact, compress, flatten, and knock out any air pockets. Tamping ensures a surface, such as sand or concrete, is even and level.

TENDER ANNUALS
These are often hot-climate or tropical annual plants that will die back completely with even a touch of frost.

UMBEL
This flower shape looks a bit like an umbrella with a number of short flower stalks coming out from a common point (like the ribs of an umbrella). They can be flat-topped or almost spherical. Classic umbels include cow parsley, fennel, and dill.

BASIC BUILDING MATERIALS

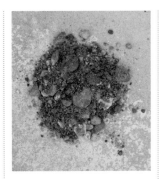

CEMENT

This powdery binding agent, when mixed with sand or ballast and water, sets to form a strong, durable material. Cement mixed with sand and water makes mortar (see below); if mixed with ballast and water, it makes concrete.

When mixing mortar for laying paving, use a mortar mix of 1:6 cement to sharp sand; for walls, you'll need something in the order of 1:3 to 1:5 cement to soft sand.

Cement comes in different colors from a really dark gray to a pale gray. Choose a color that ties in with your stone or brickwork and that sits well with the local architectural style.

SAND

When it comes to building projects, there is more than one type of sand. The different sands have quite particular uses, so don't be tempted to substitute one for another.

Soft sand, also known as builder's sand or bricklaying sand, is used in mortar mixes (see below) for walls and steps, as well as for pointing brickwork and rendering.

Sharp sand—a coarser and grittier sand—is used for mortar mixes for bedding in slabs or brick paving or underneath brick edging.

Fine kiln-dried sand can also be used to fill the gaps in brick paving.

BALLAST

A type of coarse aggregate, ballast is made up of a mixture of sand and gravel and a big mix of smaller and larger particles. When combined with cement—which works as a kind of glue—all the particles bind together to form concrete, which is strong and durable.

As with mortar mixes (see opposite), the proportion of cement to ballast that makes up concrete can vary anywhere from 1:6 to 1:12 cement to ballast, depending on what you are using it for.

BASE ROCK

You use base rock to create a solid base on which to construct paths, driveways, and load-bearing floors (for instance, a terrace). Once it's laid, you will need to compact it to create a solid, even layer on which to lay your finished surface.

This aggregate—also known as MOT Type 1, Type 1, road planings, and subbase—is made from construction waste, crushed quarry waste, rocks, and gravel.

The composition of base rock varies from region to region and also comes in different grades: the cleaner the grade, the more expensive it tends to be.

THE RIGHT MORTAR FOR THE JOB

Mortar—a mix of cement and sand—has varied mixes for various uses. If you're doing vertical work, such as building walls and steps, or for rendering and pointing, you'll need to mix cement with fine-graded soft sand in a 1:4 ratio. If you're doing what's known as horizontal work—bedding in slabs or brick paving or underneath brick edging—then you can mix cement with sharp sand in a 1:6 to 1:8 ratio.

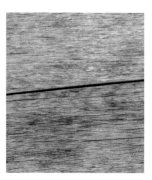

GRAVEL

A loose, decorative aggregate, gravel is made from stone fragments and classified according to the range of particle sizes. There are two main types: clean gravel and self-binding gravel.

Clean gravel can come from a gravel pit or riverbed or can be manufactured by quarrying and crushing sandstone, limestone, or granite (pea gravel is an example of clean gravel, made up of small, smooth, rounded stones).

Self-binding gravel is a mixture of variously sized gravel graded from larger particles to fine material. It forms a solid surface when compacted.

BRICKS

Although most bricks are made of clay or concrete, they come in myriad colors, finishes, textures, and sizes—as well as a massive range of prices. The standard size is 8½in x 4in x 2½in (21.5cm x 10.25cm x 6.5cm). Some have a flat surface, while others have a depression (known as a "frog"). Bricks being laid on the ground (for example, a path) need to be frost-resistant. Facing bricks are the ones commonly used to build garden walls; cheaper common bricks or concrete bricks are useful in the back course of a wall as they aren't visible. I tend to use engineering bricks for anything underground, as they are strong and particularly frost-resistant.

Bricks arrive on palettes. Since batches of bricks can vary in color and texture, always use bricks from all batches to get a good mix.

PAVING

There's a lot of paving options out there, but the biggest difference is between natural stone and concrete.

Different regions have different types of natural stone paving, depending on the region's geology. Prices will vary enormously depending on whether you buy locally or import from overseas. Always buy stone from a reputable supplier who has sourced materials ethically. Remember, because it is a natural material, stone slabs laid outdoors will react to the elements and behave differently to the same stone laid indoors.

Concrete products tend to be more economical than stone.

Both natural stone and concrete paving slabs can be purchased in varying sizes, most will be priced by the square yard (meter).

WOOD

There are two main types of wood available to use in garden builds: hardwood and softwood (see pp58–59). Hardwood is the more expensive of the two and gets more beautiful with age. That said, softwood can look great, too, especially if you paint or stain it.

You can buy hardwoods as "green" (recently cut) or "dried" (kiln-dried, air-dried, or age-dried). Expect green wood to twist and bend a bit more over time than dried wood.

A lot of wood used for construction is softwood. To make sure it doesn't rot, it will be pressure treated or tanalized (available as "green" or "brown").

When buying wood, always check that it is reasonably straight, isn't cracked too much, and doesn't have too many knots in it. Use a reputable supplier so that you know the wood is ethically sourced.

INDEX

ACKNOWLEDGMENTS

PICTURE CREDITS

The publisher would like to thank the following for their kind permission to reproduce their photographs:

(Key: a-above; b-below/bottom; c-center; f-far; l-left; r-right; t-top)

123RF.com: Petr Baumann 112br, Tatiana Belova 234c, Elena Burditckaia 59bc, claudiodivizia 59br, Stefano Clemente 33, Nataliia Kolomeitseva 27clb, lianem 113bc, Izfizf 98bc, mysikrysa 15c, Andrew Oxley 232–233, Radka Palenikova 107cr, Gerald Reindl 124bc, rootstocks 99bl, Dmytro Tolmachov 214tl, 224tl, Yoshie Uchida 120tl, Iva Vágnerová 88tr, Birute Vijeikiene 103bc, 210tl, Natasha Walton 100–101r, 116c; **Alamy Stock Photo:** A Garden 65b, A.D.Fletcher 95br, age fotostock 104bl, Avalon / Photoshot License 93r, 94bc, 97c, 125tl, Bloom Pictures 238–239, Tomek Ciesielski 152–153, Joel Douillet 106bl, RM Floral 97tl, Florapix 98cr, Tim Gainey 107clb, Garden World Images Ltd 44–45, 98cl, 100br, 104bc, 105br, 106tr, 113cl, 124c, 125bl, 125br, GFK-Flora 102bl, John Glover 98tr, 100bc, 121bl, Michele and Tom Grimm 48br, Steffen Hauser / botanikfoto 236–237, Frank Hecker 95tr, Miriam Heppell 4tr, 224–225, 226–227, Holmes Garden Photos 105tl, imageBROKER 108bl, 112tl, 222–223, Martin Hughes-Jones 101cr, 116br, LianeM 113tl, John Martin 99bc, 113clb, mauritius images GmbH 59tc, Rex May 100tl, 102br, 120cl, 125tc, 125tr, 226bl, Gerry McLaughlin 107bl, Michael David Murphy 108bc, Malcolm Park / Alamy Live News 36–37, Jacky Parker 105tr, Jaime Pharr 105bl, Pix 109cr, John Richmond 101bc, Margaret Welby 105bc; **Depositphotos Inc:** simoneandress 106br, Vilor 106clb; **Dorling Kindersley:** Zia Allaway / RHS Chelsea Flower Show 2012 99cr, 113tr, Peter Anderson 101br, 117tr, 108tr, Peter Anderson / RHS Chelsea Flower Show 2011 30–31, Peter Anderson / RHS Hampton Court Flower Show 2014 48tr, Mockford and Bonetti / Fondazione Bioparco di Roma 82br, Brian North / Waterperry Gardens 126b, RHS Tatton Park 228c, 123RF.com: Tania Sohlman / taina 14bl, 123RF.com / Veronika Surovtseva / surovtseva 59tr, Mark Winwood / Dr Mackenzie 218–219, Mark Winwood / Hampton Court Flower Show 2014 27br, 107crb, Mark Winwood / RHS Chelsea Flower Show 2014 117cr, Mark Winwood / RHS Wisley 27bl, 97bc, 98br, 101tc, 102tr, 102bc, 103c, 103cr, 104tr, 113crb, 117tl, 120br, 124bl, 212tc, 219tc, 219tr, 228-229, 234–235; **Dreamstime.com:** © Peregrine 94tr, 116tl, © Christian Weiß 103tl, Wiertn 109c; **Adam Frost:** 2, 8bl, 8br, 9tl, 9tr, 9bl, 9bc, 9br, 10–11, 12tl, 12b, 14c, 15b, 29, 32t, 32b, 38tr, 38b, 40t, 40b, 46t, 48bl, 50t, 54–55, 58t, 58bc, 60, 61, 64, 65t, 66t, 68, 69, 70br, 71, 71t, 73t, 74, 75, 76b, 77b, 79t, 79bl, 80tr, 80br, 81, 86–87, 88, 90, 91t, 91bl, 97cr; **GAP Photos:** Elke Borkowski. Design:Robert Myers 73b, Stephen Studd. Designer: Adam Frost. Sponsor: Homebase 130–131; **Garden World Images:** John Martin 99tl; **The Garden Collection:** FP / Purta 108br; **Getty Images:** L Alfonse 70bl, Anne Green-Armytage 120bc, Chris Burrows 106cr, CatLane 214br, Christopher Fairweather 97cl, Kate Gadsby 100bl, Masafumi Kimura / a. collectionRF 95tl, Ryan McVay 77t, Clive Nichols 70t, 91br, Photographed by MR.ANUJAK JAIMOOK 95bc, Carol Sharp 107br, Frank Sommariva 94bl; **Jason Ingram:** 8bc, 38tl, 211bc; **Rex by Shutterstock:** Bob Gibbons / Flpa / imageBROKER 101bl, 117bc; **Shutterstock:** Bernd Schmidt 109tl; **SuperStock:** Eye Ubiquitous 45tr, 47, 71b, 78b.

Cover images: *Front:* **Dorling Kindersley:** Peter Anderson / Adam Frost / RHS and garden t; **Adam Frost:** crb; **Jason Ingram:** cb, bl; *Back:* **Alamy Stock Photo:** A Garden bc; **Adam Frost:** clb; **GAP Photos:** Heather Edwards tr; **Jason Ingram:** crb; *Spine:* **GAP Photos:** Elke Borkowski / Adam Frost t

All other images © Dorling Kindersley
For further information see: www.dkimages.com

FROM ADAM

It still feels a little strange as to where life is taking me, but none of it would have happened without the people I have around me, and this book is no different! First, the team at DK; it feels wrong to pull out one name above another. So a massive thank you to you all for your dedication. Also thanks to Juliet Roberts, it's been fun! Thank you to the RHS, especially Sue Biggs, Chris Young, and Rae Spencer-Jones for the years of support. The photos in the book really help tell a story. A few are mine, others are my old mate Tats', but most are the work of Jason Ingram and Sarah Cuttle—you two are really top-draw. My team at home, you're all stars. Mike and Will for helping me with the projects. Nicola, Polly, Jane, and Babs, you have all brought something to the party, thank you xxxxx. And finally, Mrs. Frost and the kids, Abbie-Jade (Babs), Jacob, Amber-Lily, and Oakley (hope you like the pics, pal), just for being my reason. Lots of love xxxxx. If I have missed anybody, please forgive me x. Finally, thank you to anyone who reads the book and all your continued support. xx Adam.

FROM THE PUBLISHER

DK would like to thank everyone in Adam's team, including Nicola Oakey, Jane Adams, and Polly Hindmarch, for all their behind-the-scenes help creating the book. Thanks also to Jason Ingram and Sarah Cuttle for photography; Sarah Hopper for picture research; Steve Crozier for retouching work; Simon Maughan for reviewing the text; Oreolu Grillo, Poppy Blakiston Houston, and Lucy Philpott for editorial assistance; Sara Robin and Sophie State for design assistance; Vanessa Bird for indexing; and John Tolluck for consulting on the US edition.

Senior editor Juliet Roberts
Editor Amy Slack
Editorial team Nikki Sims, Dawn Titmus, Constance Novis
US editor Jennette ElNaggar
Senior designer Barbara Zúñiga
Designers Harriet Yeomans, Hannah Moore, Mandy Earey, Vicky Read
Jacket designer Nicola Powling
Jackets co-ordinator Lucy Philpott
Senior producer, pre-production Tony Phipps
Producer Igrain Roberts
Creative technical support Sonia Charbonnier
Managing editor Stephanie Farrow
Managing art editor Christine Keilty
Art director Maxine Pedliham
Publisher Mary-Clare Jerram

Photographers Jason Ingram, Sarah Cuttle

First American Edition, 2020
Published in the United States by DK Publishing
1450 Broadway, Suite 801, New York, NY 10018

A catalog record for this book is available from the Library of Congress.
ISBN 978-1-4654-7285-4

DK books are available at special discounts when purchased in bulk for sales promotions, premiums, fund-raising, or educational use. For details, contact: DK Publishing Special Markets, 1450 Broadway, Suite 801, New York, NY 10018
SpecialSales@dk.com

Printed and bound in China

A WORLD OF IDEAS:
SEE ALL THERE IS TO KNOW

www.dk.com

'To my old man, this one is for you. I'm on my way!'

ABOUT THE AUTHOR

Adam Frost is an award-winning garden designer, with seven Royal Horticultural Society (RHS) Chelsea Flower Show Gold Medals to his name. He is a presenter on BBC *Gardeners' World* and BBC coverage of the RHS Flower Shows.

He started his gardening career as a 16-year-old gardening apprentice with the North Devon Parks Department. He then moved back to London to train as a landscape gardener, before getting a job with the late great gardener Geoff Hamilton. During his time with Geoff, he trained as a landscape designer and then established his own garden landscape business in 1996, which has taken him around the world designing gardens.

In 2017, he set up The Adam Frost Garden School at his home in Lincolnshire where he hosts informal workshops to gardening enthusiasts.

Adam is also an RHS Ambassador, which has a mission close to his heart: to encourage our next generation of gardeners and to raise the profile of careers in horticulture.

Adam and his wife, Sulina, live in a picturesque Lincolnshire village with their 4 children, 1 horse, 2 ponies, 2 dogs, and an ancient cat! His 5-acre garden is regularly featured on *Gardeners' World* as a work in progress.